IGNITING
INNOVATION

IGNITING

INNOVATION

Rethinking the Role of Government in Emerging Europe and Central Asia

Itzhak Goldberg
John Gabriel Goddard
Smita Kuriakose
Jean-Louis Racine

Europe and Central Asia Region

THE WORLD BANK

©2011 The International Bank for Reconstruction and Development / The World Bank
1818 H Street NW
Washington DC 20433
Telephone: 202-473-1000
Internet: www.worldbank.org

1 2 3 4 14 13 12 11

This volume is a product of the staff of the International Bank for Reconstruction and Development / The World Bank. The findings, interpretations, and conclusions expressed in this volume do not necessarily reflect the views of the Executive Directors of The World Bank or the governments they represent.

The World Bank does not guarantee the accuracy of the data included in this work. The boundaries, colors, denominations, and other information shown on any map in this work do not imply any judgement on the part of The World Bank concerning the legal status of any territory or the endorsement or acceptance of such boundaries.

ISBN: 978-0-8213-8740-5
e-ISBN: 978-0-8213-8741-2
DOI: 10.1596/978-0-8213-8740-5

Cover illustration & design: Romain Falloux
Cover image of Sputnik-1: NASA/Asif A. Siddiqi

Library of Congress Cataloging-in-Publication Data
   Igniting innovation : rethinking the role of government in emerging Europe and Central Asia / Itzhak Goldberg ... [et al.].
      p. cm.
   Includes bibliographical references.
   ISBN 978-0-8213-8740-5 (alk. paper)
   ISBN 978-0-8213-8741-2 (e-ISBN)
      1. Research, Industrial—Economic aspects. 2. Endowment of research—Europe. 3. Endowment of research—Asia. 4. Government spending policy. I. Goldberg, Itzhak.
   HC79.R4I38 2011
   338′.064094—dc23
                        2011024386

Contents

Foreword   *ix*
Acknowledgments   *xiii*
Contributors   *xv*
Abbreviations   *xvii*

Overview   **1**
Why innovation matters   4
Acquiring technology from abroad   9
Connecting research to firms   12
Restructuring options for RDIs   14
Bringing innovations to market   14

**1. Why innovation matters—and what the government
should do about it**   **19**
The rationale for innovation   26
Coping with spillovers   28
Coping with unequal information and the "funding gap"   30
Why the government should play a role   33

**2. Acquiring technology from abroad—leveraging the
resources of foreign investors and inventors**   **39**
Cross-border knowledge flows   42
Acquiring foreign technology   54
How well do ECA firms absorb knowledge?   62
Case study: The role of FDI in helping Serbia acquire
technology   69

**3. Connecting research to firms—options for reforming the
     public RDIs**   **81**

Incomplete restructuring of RDIs   83
RDIs outside ECA   85
A snapshot of RDIs in ECA   91
Government funding and governance   100
A roster of obstacles   106
A proposed RDI reform strategy   110
Options for restructuring RDIs   112
Options for public funding to support RDI reforms   115
Case Study: Finland's shift to a knowledge-based economy:
     The Role of TEKES   118

**4. Bringing innovations to market—boosting private
     incentives through public instruments**   **123**

Basic principles of instrument design   125
Basic types of instruments   131
Financial instruments for ECA   135
Institutional support instruments   144
Monitoring and evaluation   148
Conclusion   151
Case study: How Israel has promoted innovation
     in recent decades   152

References   **157**

Boxes

1   Demystifying innovation and absorption   5
2   Poland at a crossroads: Expanding from technology
     absorption to broader meanings of innovation   6
1.1 Defining innovation and absorption   21
2.1 A snapshot of coinvention in Poland   53
3.1 Restructuring of RDIs faces important legacy challenges   109
4.1 Catalyzing private sector innovation in Turkey through
     an improved institutional environment and financial
     instruments   127
4.2 Using grants and loans for innovation support in Croatia   132
4.3 Do matching grants for industrial R&D help the Israeli
     economy?   138
4.4 Armenia's efforts at enterprise incubation   144

Figures

1	ECA's R&D efficiency is still low	8
2	ECA needs to boost its R&D spending	8
3	The expanding role of international coinvention in the ECA 7	10
4	A call for policy reforms and capacity building	13
1.1	ECA needs to boost its R&D spending	23
1.2	ECA's researcher population is unevenly distributed	24
1.3	ECA's R&D efficiency is still low	25
1.4	Corporate ventures and the government play a key role in the early stages	32
2.1	Innovation and absorption spur growth and productivity	41
2.2	ECA inventive activity on the rise	44
2.3	Hungary and the Czech Republic lead the ECA patents race	45
2.4	Russian Federation's patent share could be even bigger given its size and scientific strength	46
2.5	EU12 losing its edge on China and India	47
2.6	The expanding role of international coinvention in the ECA 7	48
2.7	Germany dominates ECA coinventions	49
2.8	Expanding role of international coinventions in the Russian Federation	50
2.9	Revenue and employment trends pre- and post-acquisition	72
3.1	Number of annual publications per RDI staff	97
3.2	R&D and technical services to industry mostly marginal compared with public funds	98
3.3	Some ECA RDIs generate as much industry revenue as international benchmarks, but this is not the norm	99
3.4	SMEs could make greater use of RDIs	100
3.5	A bias toward a few types of funding sources	101
3.6	Too few private sector board directors	103
3.7	RDIs' salaries not attractive enough	106
3.8	Chain of events leading to ineffective RDIs	108
3.9	Factors affecting RDI performance	108
3.10	RDI restructuring strategies	111
3.11	RDI reform decision tree	112
3.12	The less the government funding, the more market pull dominates	116
3.13	Finland's business sector is sharply stepping up its R&D	119
3.14	Finland's successful innovation environment	120

Tables

2.1   Top generators of Russia-based U.S. patents   51

2.2   Openness is better: Link between international
      interconnectedness and technology absorption   65

3.1   A massive overhaul of RDIs in the 1990s   84

3.2   A successful strategy typically reflects market needs   88

3.3   Foreign comparator RDIs vary in size and ownership   93

3.4   Specializations of the RDIs in the ECA sample   94

3.5   Restructuring options for ECA RDIs   113

Foreword

Innovation and technology absorption are now firmly recognized as one of the main sources of economic growth for emerging and advanced economies alike. That is why igniting the latent potential of the innovation systems in Europe and Central Asia (ECA) is seen as a possible catalyst for revitalizing the economies after the recent shocks of the global financial and economic crisis. More active government intervention to support innovation is being weighed as a tool to transform failing industries, develop new industries, and speed up a recovery process of export-oriented industries. Even for countries lucky enough to have substantial natural resources, there is a pressing need to move up the value chain and diversify their economies to mitigate future price shocks.

Is there a role for government intervention to ignite innovation in ECA? This is the central question to which this book responds. The answer is yes, but a qualified yes. Innovation activities are rife with market failures that tend to hold back private investment. Yet at the same time badly designed or badly implemented interventions can further hamper the development of an innovative and entrepreneurial culture among businesses and research communities. The new evidence marshaled by this book coincides with a growing academic literature that shows that interventions will lead to firm-level productivity and sustainable growth under certain conditions—domestic competition, international trade flows, research and development (R&D) collaboration, worker mobility, foreign direct investment (FDI), and good governance and transparency in the innovation agencies.

For ECA, a first step in revitalizing its economies will be absorbing global technology through trade, FDI, or licensing. But given that this process is not guaranteed or cost free, firms and countries will need to invest in developing their absorptive capacity. The good news is that there is a tradition of excellence in learning and basic research in many ECA countries that provides the base for future commercial innovation. But going from a strong research foundation to economically productive commercial applications remains a critical missing link in ECA countries. Two decades after the transition, many of the public R&D institutes are still operating as standalone entities with limited progress in the intensity and quality of their interactions within the national innovation system and, specifically, services they provide to industry. They constitute a legacy of central planning that has yet to be fully resolved. One of the contributions of this book is to point out potential strategies to restructure public R&D institutes with the aim of effectively aligning their activities with private efforts. The restructuring of the public R&D institutes will rectify the balance between private and public R&D and ensure that the institutes can play a real role in igniting innovation-led growth.

This book builds on the lessons from public institutions and programs to support innovation, both successful and failed, from ECA as well as China, Finland, Israel, and the United States. Field visits to these countries were hosted by the innovation and scientific agencies of the respective governments, strengthening the international experiences presented here. The lessons highlight the pitfalls of imitating models of government interventions from "innovative" countries without having adequate systemic governance and institutional reforms. They underscore the need for intensified international R&D collaboration and foreign R&D investment to better integrate ECA in the global R&D community. They spotlight further opening to FDI to encourage knowledge absorption. And they point to the importance of overhauling government support programs—especially financial ones—to address key pressures points along the innovation and commercialization continuum.

This book is a culmination of 10 years of analytic and operational work led by the Private and Financial Sector Development Department and the Chief Economist's Office of the ECA region of the World Bank. Several regional reports and country policy notes exploring these issues have been published over the years. The book also reflects the lively discussion in the ongoing series of flagship events to promote knowledge-based economies in the region. The most recent Knowledge Economy Forum was held in Berlin in 2010, hosted by the Fraunhofer Center for Central and Eastern Europe. The Forum has traveled to many parts of the ECA region, as well as France, the United Kingdom, and other European Union countries.

The title of the book, *Igniting Innovation: Rethinking the Role of Government in Emerging Europe and Central Asia*, reflects our belief in the central impact of the public sector on innovation. The book identifies policies that have an adverse affect on innovation. It also identifies policy gaps, that, if filled, could have a catalytic effect on private sector innovation. There are many new ideas and technologies coming out of the ECA region—for example Skype, the work of Estonian software developers in partnership with Scandinavian entrepreneurs. We hope that the results and recommendations offered by this book will contribute to the discussion about how to ignite innovation and technology adoption as a central part of the development and growth strategies of ECA countries.

Philippe Le Houérou   Gerardo Corrochano
Vice President   Director, Private and Financial Sector
Europe and Central Asia Region   Europe and Central Asia Region
The World Bank   The World Bank

Acknowledgments

This book has been sponsored by the Chief Economist's Office in the Europe and Central Asia Region of the World Bank. The book was prepared by a team in the Finance and Private Sector Development unit of the Europe and Central Asia region consisting of Itzhak Goldberg, John Gabriel Goddard, Smita Kuriakose, and Jean-Louis Racine.

The team would like to recognize the coauthors of three previous regional reports, which served as the foundation for this book and helped to guide the team's work on this subject. They include Manuel Trajtenberg (Professor at the Eitan Berglass School of Economics, Tel-Aviv University); Adam Jaffe (Dean of Arts and Sciences and Fred C. Hecht Professor in Economics, Brandeis University); Julie Sunderland (Oriane Consulting); Thomas Muller and Enrique Blanco Armas (both World Bank) for ECA Knowledge Economy Study I, "Public Financial Support for Commercial Innovation"; Lee Branstetter (Professor, Heinz School, Carnegie Mellon University) for ECA Knowledge Economy Study II, "Globalization and Technology Absorption in Europe and Central Asia: The Role of Trade, FDI and Cross-border Knowledge Flows"; and Natasha Kapil (World Bank) for ECA Knowledge Economy Study III, "Restructuring of Research and Development Institutes in Europe and Central Asia."

The team would also like to thank Bruce Ross-Larson, Laura Wallace and Jack Harlow, who provided vital editorial support throughout the entire project. Paola Scalabrin, Dina Towbin, and Aziz Gokdemir provided support with the production, and Romain Falloux of El Vikingo Design, Inc. designed the book. We are also grateful to Daniel Lim for written contributions and Christina Tippman and Yana Ukhaneva for research assistance on the book and acknowledge inputs received from Yesim Akcollu, Paulo Correa, Naoto Kanehira, Natasha Kapil, Marcin Piatkowski, Sameer Shukla (all World Bank) and Bagrat Yengibaryan,

(Director, Enterprise Incubator Foundation, Armenia). We acknowledge significant contributions received from Ali Mansoor and Mohammed Ismail for ECA Knowledge Economy Study I, "Public Financial Support for Commercial Innovation"; from Martina Kobilicova, Andrej Popovic, Lazar Sestovic, Ana Margarida Fernandes, Paulo Guilherme Correa, Mallika Shakya, Chris Uregian, and Jasna Vukoje for Knowledge Economy Study II, "Globalization and Technology Absorption in Europe and Central Asia: The Role of Trade, FDI and Cross-border Knowledge Flows"; and from Irina Dezhina, Roberto Mazzoleni, Joanna Tobiason, Slavo Radosevic, and Gilbert Nicolaon for ECA Knowledge Economy Study III, "Restructuring of Research and Development Institutes in Europe and Central Asia."

The team would like to acknowledge valuable guidance received on the evolving manuscript from peer reviewers Mark Dutz, William Maloney, and Milan Brahmbhatt. Valuable written comments were also received from Lilia Buruncic (Country Manager, Macedonia) and Zeljko Bogetic (Lead Economist for Russia).

The team especially thanks Gerardo Corrochano (Director of the Financial and Private Sector Development Department, Europe and Central Asia Region and Director of the Innovation, Technology and Entrepreneurship Practice, Financial and Private Sector Development [FPD] Network) and Fernando Montes-Negret (former Director of the Financial and Private Sector Development Department, Europe and Central Asia Region), for their continued support, comments, and guidance for this book.

The team appreciates the intellectual leadership and support received from Indermit Gill (Chief Economist, Europe and Central Asia) and Pradeep Mitra (formerly Chief Economist, Europe and Central Asia), as well as Marianne Fay (Chief Economist of the Sustainable Development Network, formerly Lead Economist in the Chief Economist's Office of Europe and Central Asia) and Willem van Eeghen (Lead Economist, Chief Economist's Office of Europe and Central Asia).

Contributors

Itzhak Goldberg is currently a Senior Researcher at the Institute for Prospective Technological Studies of the European Commission's Joint Research Centre and a Fellow at CASE – Center for Social and Economic Research, Warsaw. He worked as an Advisor for Policy and Strategy with the Europe and Central Asia Region of the World Bank from 1990 to 2009, where he was in charge of private sector development programs in various countries in the region, and he was pivotal in the design and implementation of the Privatization Program of the Government of Serbia. More recently, he has devoted his attention to the economics of innovation and technology upgrading in the region, leading the series of studies on Knowledge Economy. His recent book (with Raj Desai) *Can Russia Compete?* was published in June 2008 by the Brookings Institution. Prior to joining the World Bank, he was the Chief Economist and member of the executive management board of Dead Sea Works Limited in Israel and Adjunct Associate Professor of Economics in the Ben Gurion University in Israel. He also worked as a Research Fellow in Hoover Institution in the United States in the late 1970s. He obtained his PhD from the University of Chicago in 1976, after studying Economics in the Hebrew University of Jerusalem.

John Gabriel Goddard is an Economist in the World Bank's Finance and Private Sector Development unit of Europe and Central Asia, where he undertakes advisory and lending operations that aim to raise the competitiveness and growth potential of countries through R&D and innovation programs, business environment improvements, and access to finance facilities. He currently coordinates the private sector development programs in Bulgaria and Uzbekistan, and since joining the World Bank in 2006, he has also taken part in the preparation and supervision of investment operations in Colombia, Croatia, Nigeria, and Uzbekistan. Prior to joining the World Bank, he worked in research positions at the Paris Dauphine University and the Cournot Centre in Paris, and as an economist in U.K.-based consulting firms and the Mexican government. He holds a bachelor's degree in Economics from CIDE in Mexico City, a master's and a doctorate in Economics from the University of Oxford, and he has been a visiting fellow at Stanford and Harvard universities.

Smita Kuriakose is an Economist in the Finance and Private Sector Development Department in the Africa Region in the World Bank. In particular, her expertise is in science, technology, innovation policy, and skills development issues with the view to promote private sector development in Africa. Smita has been a key team member in the ongoing skills and innovation studies in Mauritius and South Africa and has coauthored the forthcoming regional study on trade, foreign direct investment, and technology absorption in Southern Africa. In addition, she has worked on lending operations in Ethiopia, Lesotho, and Mauritius. Previously, she worked on innovation and investment climate issues in the Europe and Central Asia region, where she coauthored regional knowledge economy studies. Prior to joining the Bank in Washington, DC, she worked on the United Nations Link Project and the PREM Unit of the World Bank in India where she focused on fiscal and macro policies. Smita holds graduate degrees in Economics from the University of Maryland, College Park, and the Delhi School of Economics in India.

Jean-Louis Racine is an Innovation Specialist in the Europe and Central Asia region of the World Bank. At the World Bank, his work focuses on policies and programs to support industrial upgrading, technology diffusion, and innovation. His work has ranged from improving the effectiveness of the broader innovation policy and governance frameworks, to restructuring of R&D institutes, and to designing programs to stimulate innovative entrepreneurship. Quality upgrading has played a central role in his approach to industrial development, having worked on transitioning countries to internationally accepted standards. In this field, he has coauthored books on Latin America and on Eastern Europe and Central Asia. Prior to his current position, he worked in a private consulting practice where he advised regional governments and businesses on technology-based competitiveness strategies. He draws from a combined background in engineering and policy, with a PhD in Mechanical Engineering from the University of California at Berkeley, an MIA in Technology Policy for Economic Development from Columbia University, and an MSc in Mechanical Engineering from Stanford University.

Abbreviations

BEEPS	Business Environment and Enterprise Performance Survey
CSIR	Council of Scientific and Industrial Research
ECA	Europe and Central Asia
EPO	European Patent Office
ESTD	Early stage technological development
EU	European Union
EU8	Czech Republic, Estonia, Latvia, Lithuania, Hungary, Poland, Slovak Republic, Slovenia
EU10	Bulgaria, Czech Republic, Estonia, Hungary, Latvia, Lithuania, Poland, Romania, Slovak Republic, and Slovenia
EU15	Austria, Belgium, Denmark, Finland, France, Germany, Greece, Ireland, Italy, Luxemburg, Netherlands, Portugal, Spain, Sweden, and the United Kingdom
FDI	Foreign direct investment
GDP	Gross domestic product
GOCO	Government-owned, contractor operated
GOGO	Government-owned, government operated
HKPC	Hong Kong Productivity Council
ICT	Information and communication technology
ITRI	Industrial Technology Research Institute
KIST	Korea Institute of Science and Technology
M&A	Mergers and acquisitions
M&E	Monitoring and evaluation
MNE	Multinational enterprise
OCS	Office of the Chief Scientist

OECD	Organisation for Economic Co-operation and Development
R&D	Research and development
RDIs	Research and development institutes
S&T	Science and technology
SAR	Special administrative region
SBIR	Small Business Innovation Research
SINTEF	Norwegian Foundation for Scientific and Industrial Research
SITRA	Finnish National Fund for Research and Development
SMEs	Small and medium enterprises
Tekes	National Technology Agency of Finland
TFP	Total factor productivity
TIP	Technology Incubators program
TNO	Netherlands Organization for Applied Scientific Research
TTO	Technology Transfer Office
UNESCO	United Nations Educational, Scientific and Cultural Organization
USPTO	U.S. Patent and Trademark Office
VTT	Technical Research Centre of Finland

Overview

"When we look at the regions of the world that are, or are emerging as, the great hubs of entrepreneurial activity—places such as Silicon Valley, Singapore, Tel Aviv, Bangalore, and Guangdong and Zhejiang provinces—the stamp of the public sector is unmistakable. Enlightened government intervention played a key role in creating each of these regions. But for each effective government intervention, there have been dozens, even hundreds, of failures, where substantial public expenditures bore no fruit."

—Josh Lerner in *Boulevard of Broken Dreams: Why Public Efforts to Boost Entrepreneurship and Venture Capital Have Failed—and What to Do About It*, 2009.

In the chapters to come, we will encounter again and again the dilemma facing policymakers dealing with public goods such as innovation and technology flows: countries cannot do without government intervention, but with government intervention, the probability of government failure is high because it is so tough to get the design and implementation right.

Yet in the wake of the recent financial and economic crisis, there have been renewed calls for tackling the market failures that are holding back private investment through government intervention. Frequently ignored are the classic examples of government failures: risks of non-transparent funding decisions and targeting support to maintain employment. In transition economies, the legacy of past government interventions that created state research and development institutes—a component of the central planning system, which conducted research and development (R&D) in isolation from enterprises—continue to "crowd out" startups and spinoffs.

At the same time, Europe and Central Asia (ECA) countries are searching for avenues to revitalize their economies. Countries that anchored their growth prospects on the financial sector, real estate, and construction are now trying to shift to a more balanced growth path—in which export-oriented industries play a larger role and a bigger share of foreign direct investment (FDI) goes to manufacturing. And countries that depend on natural resources are looking to move up the value chain as well as diversify their economies.

Technology adoption and innovation can be a catalyst for this revitalization. Given the likely decline in the potential growth rate in ECA countries in the medium term, returning to precrisis GDP growth rates will require faster productivity growth. An abundant literature emphasizes the close link between technology absorption and innovation and long-term, self-sustained economic growth. As a result of suboptimal spending on R&D and structural weaknesses in the national innovation system, growth in total factor productivity, which usually represents more than half of long-term economic growth across countries, may be undermined.

Two candidates for this type of revitalization include the early reformers of Central and Eastern Europe; the EU10 countries—Bulgaria, Czech Republic, Estonia, Hungary, Latvia, Lithuania, Poland, Romania, the Slovak Republic, and Slovenia—which joined the European Union (EU) in 2004–07; and a large resource-rich country, the Russian Federation.

For the EU10, its fate is wrapped up with the EU, whose EU 2020 strategy presents a coordinated response to today's economic challenges. The strategy focuses on making the economies more productive through structural reforms and a renewed emphasis on knowledge and innovation. The hope is that smart, sustainable, and inclusive growth will deliver high levels of employment, productivity, and social cohesion—especially at a time of aging populations and stepped up competition from developed and emerging economies, such as China and India.

The strategy confirms and updates the Lisbon Agenda targets agreed in 2000, including the objective of ramping up private and public investment in R&D to 3 percent of GDP for the EU as a whole, but now also setting more realistic national targets. Achieving these goals will require concerted and sustained action by European governments. But even this may not be enough, unless the instruments used to intervene and influence the behavior of knowledge providers—such as universities, research development institutes (RDIs), and knowledge users in the private sector—are improved, especially so that R&D results in the public sphere are implemented by enterprises in a more systematic approach.

For Russia, its dependence on oil and gas exports has significantly risen in the past decade resulting in rising fiscal risks to oil price volatility (Bogetic and others 2010). Meanwhile, the fragility of the competitiveness of Russian enterprises has become increasingly visible with global competition and the greater competitiveness of Brazil, China, India, and other Asian countries. Stimulating economic diversification can cover a wide number of issues and involve many challenges, including entrepreneurship, foreign investment, regional development, and physical infrastructure. In Russia's case, it comes down to one thing: ensuring that the manufacturing sector can compete in the global economy. Russia has come a long way over the past decade. The World Bank's analyses of short-term growth (*Russian Economic Reports Series*, for example World Bank 2011) and long-term growth (World Bank 2005b) and productivity (Alam and others 2008) note the key role of prudent macro-fiscal policy (including the creation and successful operation of the oil funds) and productivity gains, which underpinned strong growth performance. But for the good times that Russia's economy has enjoyed to be sustained, policymakers need to address several bottlenecks that have slowed the country's transition to a knowledge-driven and entrepreneurial economy (Desai and Goldberg 2008).

Our book, titled *Igniting Innovation: Rethinking the Role of Government in Emerging Europe and Central Asia*, is a call for the ECA region to take dramatic steps to position itself closer to the scientific and technological frontiers and regain the lead in regional or even global settings. We explore why and how ECA countries should use innovation and absorption to pursue faster growth, starting with the rationale for public intervention and support. We suggest that while in most areas of economic activity market forces can—most of the time but not always as we have recently seen—ensure efficiency and social welfare, innovation also requires government intervention. However, we caution that learning from "innovative" countries (such as Finland, Israel, Republic of Korea, and Singapore) and emulating their models of government interventions require better governance and institutions than most ECA countries currently have. This means that public support could be wasted without parallel improvements in economic incentives, upgrading information infrastructure, and reforming education. Even worse, intervention without these prerequisites might cause harm as businesses might manipulate the policies or rules of the innovation support "game" to their advantage.

The key throughout is a *firm-centered* approach to technology absorption and innovation, because one of the major challenges for ECA countries is to rectify the balance between private and public R&D. To achieve a fundamental change in the culture of the business sector and firm-level incentives, it will be essential to nudge firm managers to take risks and abandon their post-transition comfortable conservatism. Firm-level incentives depend, as is vividly illustrated in our case study of FDI in Serbia, on making managers genuinely responsible to shareholders (corporate governance), removing barriers to startups and spinoffs (investment climate), and stimulating firms to invest in skills and R&D, despite the risks of labor turnover and theft of intellectual property.

Our book analyzes four aspects of the innovation system: international collaboration, research and development institutions, government financial support instruments, and the investment climate. It concludes that the following would ignite innovation in emerging Europe and Central Asia:

- **Supporting the collaboration of local researchers and foreign inventors—and attracting foreign R&D investment.** To enhance ECA's integration into the global R&D community, governments should support local coinventors in obtaining international patent protection before they negotiate the ownership of their joint patents with their western coinventors.

- **Restructuring research and development institutes.** To better commercialize R&D outputs, policymakers should focus on the real potential of each organization and on the local and global demand for their outputs—that is, on their economic viability.

- **Rethinking financial support instruments.** To promote risk-taking and stimulate markets for private risk capital, policymakers should evaluate support instruments and develop new ones based on international good practices—such as matching grants, minigrants, venture capital, innovation parks, incubators, and angel investors.

- **Facilitating trade, FDI, and entrepreneurial start-ups and spinoffs.** For reforms in innovation to make a difference, governments need to make it easier to do business in ECA in measurable ways—with a special emphasis on the needs of FDI, start-ups, and spinoffs, such as starting a business, protecting investors, and getting credit and enforcing contracts.

Why innovation matters

If ECA countries had hoped to grow through external financing or natural resource exports, the recent global financial crisis exposed the limits of growth through these means. The reality is that the scarcity of capital and high cost of labor in some ECA countries and low productivity in others, relative to Asian competitors, circumscribe technology absorption and innovation-based competitiveness.

To resolve persisting structural weaknesses, policymakers are turning to innovation and technology absorption to serve as a postcrisis source of growth. Yet the adoption of existing technology through trade, FDI, or licensing is not guaranteed or cost free. Firms and countries need to invest in developing absorptive capacity. The good news is that a tradition of excellence in learning and basic research in several ECA countries pro-

BOX 1
Demystifying innovation and absorption

- *Innovation* is the development and commercialization of products and processes that are new to the firm, new to the market, or new to the world. The activities involved range from identifying problems and generating new ideas and solutions, to implementing new solutions and diffusing new technologies.

- *Absorption, a subset of innovation,* is the application of existing technologies, processes, and products in a new environment in which they have not yet been tested and the markets and commercial applications are not fully known—that is, they are "new to the firm."

The two can work together to create a virtuous circle. Innovation promotes *absorptive capacity*—a firm's capacity to assess the value of external knowledge and technology, and make necessary investments and organizational changes to absorb and apply this knowledge and technology in its productive activities. And the economy's ability to research and develop new technologies boosts its ability to understand and apply existing technologies. Vice versa, the absorption of cutting-edge technology inspires new ideas and innovations.

Source: Authors.

vides hope that commercial innovation could be adopted and built "on the shoulders" of the past. But translating this research foundation into economically productive commercial applications remains a critical missing link in ECA countries.

R&D spending lies at the heart of absorptive capacity but tends to be lower in ECA countries, where the average R&D-to-GDP ratio is 0.9 percent, well below the world average of 2.4 percent—though there is a wide variation in the intensity of R&D spending across ECA countries. Some Nordic countries like Finland and Sweden spend close to 4 percent of GDP on R&D, around twice the Organisation of Economic Co-operation and Development (OECD) and EU27 averages. The large OECD countries, including France, Germany, the United Kingdom, and the United States have an R&D intensity of between 2 and 3 percent. The Czech Republic and Slovenia, the ECA economies with by far the higher R&D intensity, spend just about 1.5 percent of GDP. As of 2009, Poland, an EU member state, had an R&D spending of only 0.7 percent of GDP, below the ECA average (box 2).

Moreover, innovation outputs are comparatively low in ECA, even when considering the level of R&D inputs. Russia lags behind OECD and other large middle-income countries in R&D output, including a lower number of patents and scientific publications per capita. In fact, they spend more on R&D than most EU15 countries for each U.S. Patent and Trademark Office (USPTO) patent they generate (figures 1 and 2).

BOX 2
Poland at a crossroads: *Expanding from technology absorption to broader meanings of innovation*

While many European countries struggled during the global financial crisis, Poland displayed a remarkable resilience. It was the only European Union (EU) country to expand, growing 1.7 percent in 2009 compared with a decline of 4.3 percent in the EU15 countries and 3.6 percent in the EU10 region.[1] But its growth was largely thanks to capital accumulation and higher productivity based on technological absorption, rather than business research and development (R&D)—the key to future growth. Poland is now at an important crossroads: it can continue to focus on absorption at the expense of innovation or do the opposite. The reality is that it must raise the value and efficiency of public R&D spending—one of the lowest in the EU—and leverage private R&D spending to generate new technologies and transfer them by collaborating with other R&D-intensive organizations.

Polish firms invest significantly more on absorption (as shown by new machinery, equipment, and software) than most neighboring Europe and Central Asian countries, and they display little interest in financing R&D (figure below). Over time, business R&D is increasing at a rate comparable with EU peers, but it will have to grow much faster to make up for the existing gap. As of 2009, Poland's R&D spending was only 0.7 percent of GDP, well below the EU average of 2 percent. Business R&D constituted only a

Poland's private sector needs to focus more on innovation

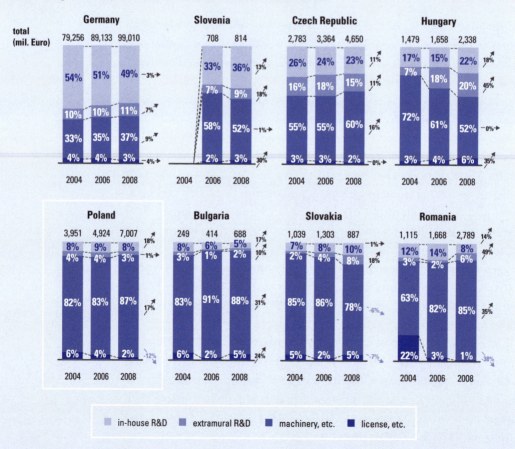

small portion of this amount, at 0.2 percent of GDP, with public R&D representing the rest.[2] In most leading European economies, the ratios are reversed.

Poland has a unique opportunity now to increase private R&D spending and boost the efficiency of public expenditures on R&D owing to its EU membership and the Structure Funds for supporting innovation. The main instrument, the Operational Programme Innovative Economy 2007–13, gives funds to Polish enterprises for raising R&D outlays, improving cooperation between the R&D sphere and businesses, and stimulating entrepreneurship. It provides €8.7 billion in funding—of which €4.1 billion directly supports enterprises and €4.6 billion does so indirectly.

To better position itself, Poland has recently adopted a comprehensive reform to improve the quality and competitiveness of science and higher education. But early estimates suggest that the bulk of the money, €2.9 billion, is supporting technology absorption, with a much smaller amount, €1.2 billion, going to innovation—through direct (matching grants, grants, loans, and venture capital) and indirect instruments (vouchers, tax incentives, and technical assistance). The problem is that public funds to cofinance technology absorption are unlikely to have the transformational impact that would occur with innovation support.

One possible explanation for the bias toward absorption is that public administrators have a low–risk appetite—preferring productivity-enhancing investments that will boost employment in the short run. But Polish policy and opinion makers increasingly agree that there will need to be a considerable change in public mindset and related risk tolerance for innovation funding. Market-based solutions could support business investments, and public funds could be redirected to support the high-risk early stages of innovation. The challenge lies in designing a mix of financial instruments that will cater to all stages of the commercialization cycle—early, growth, and mature—with an emphasis on high-tech startups that can generate novel ideas and products and catapult Poland into the global innovation league.

Notes: 1. World Bank 2011a. 2. European Commission 2011.
Sources: European Commission 2011; World Bank 2011a and b.

On top of raising public R&D and stimulating private R&D, revitalizing growth will require new strategies to raise competitiveness: export-led sectoral diversification, investments in skills and infrastructure, and commercialization of new ideas and absorption of technologies from the world. Is there a role for government intervention in these areas? Yes, because of several market failures. For example, the government can help by promoting startups to generate new activities and support sustainable job creation, with programs that mitigate the high failure risk that deters the entry of new ventures and the ceilings on growth once such ventures are established.

That said, there is reason to ask how far governments should go in their interventions. It is a timely question given that many countries are stepping up their use of industrial policy—that is, an attempt by the government to actively promote the growth of particular industrial sectors and companies, long known as "picking winners" and saving losers—despite the controversy surrounding the topic.

FIGURE 1
ECA's R&D efficiency is still low

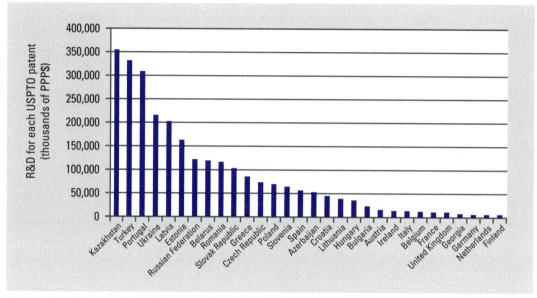

Source: Authors' calculations based on UNESCO and USPTO data for 2007.

A new approach to thinking about industrial policy promulgated by Dani Rodrik is a discovery process, one where firms and the government learn about underlying costs and opportunities and engage in strategic coordination. Given that our book focuses on the government's role in supporting innovation, the debate on industrial policy, which sets the framework for the design of support instruments, is pertinent for our analysis. A key principle in the design of support (for example, matching

FIGURE 2
ECA needs to boost its R&D spending

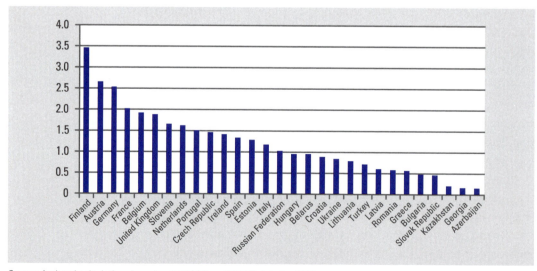

Source: Authors' calculations based on UNESCO and USPTO data for 2007.

grants) is "neutrality" with respect to sectors in project selection. Neutrality means that the selection criteria are based on the project's value (cash flow net present value) and not on the sector of the firm.

Should ECA support instruments be neutral about sectors? Two of the world's most advanced economies, the United States and the EU, apply a mix of technology-neutral and technology-specific approaches. In these economies, which have highly qualified civil servants in charge of selection, exceptions to neutrality for technologies that exploit a comparative advantage, or for technologies associated with public goods, could make sense. But some ECA governments may lack the capacity to implement non-neutral project selection—a process that requires sophisticated information—rendering them subject to enormous pressures from vested interests.

Even so, many ECA governments are weighing industrial policy. Worried about excessive reliance on extractive industries, Russia and other resource-rich ECA countries are considering industrial policies to diversify and modernize—for example, mega-projects in Russia to jumpstart a high-tech research and production hub in Skolkovo along the lines of Silicon Valley. At the same time, the number of innovative small and medium enterprises (SMEs) in Russia is small, and the budget to support innovative SMEs (such as the Bortnik Fund) is minuscule.

Acquiring technology from abroad

Improving the ability of ECA countries to tap into the global technology pool is a vital mechanism for accelerating industrial development, worker productivity, and economic growth. But the process of knowledge absorption is neither automatic nor costless.

How does absorption occur? It requires dense linkages to the global knowledge economy, human capital, and a learning-by-doing process, among other factors. Trade flows, FDI, R&D, labor mobility, and training are among the best conduits—in effect "channels of absorption." And it is often the case that extensive, active efforts are required to adapt technology pioneered outside the region—in large ways and small—to the economic circumstances of ECA countries.

This chapter analyzes the extent of knowledge and technology absorption for firms in ECA, as well as the factors that influence absorption. It asks two critical questions:

- To what extent are ECA countries able to leverage international knowledge flows and cross-national technological cooperation in the ECA region, as measured by patents and patent citations?

- What is the role of openness to trade and FDI, participation in global supply networks, and investment in human capital, through on-the-job training, as seen through the eyes of ECA manufacturing firms?

On the first question, the answer is: "not enough." Although ECA inventive activity, measured by patents, has been on the rise over the past decade, it has been concentrated in just a few of the more advanced economies. Moreover, the relative isolation of ECA R&D is underscored by the relatively small number of citations these patents receive in patents subsequently granted. In addition, foreign firms appear to be making a significant contribution to inventive activity in the ECA region. These firms' local R&D operations, and their sponsorship of local inventors, generate a large fraction of the total patents emerging from ECA countries. In ECA, our empirical work shows new evidence of the growing use of coinventions, with more than half of total patent grants generated from teams of investors in more than one country (figure 3). Inventors in western economies and Germany play a particularly important role.

This process of international coinvention not only contributes to the quantity of ECA patents but also raises the quality of ECA inventive efforts. Whereas indigenous ECA patents lag behind other regions in

FIGURE 3
The expanding role of international coinvention in the ECA 7

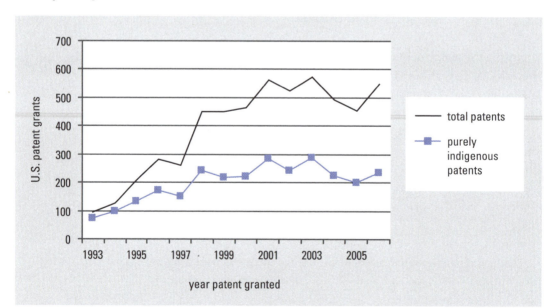

Note: The graph tracks total counts of patents in which at least one inventor is based in one of seven ECA countries: Bulgaria, the Czech Republic, Hungary, Poland, the Russian Federation, Slovenia, and Ukraine. "Purely indigenous patents" are those generated by a team whose members are all based in a single ECA country.
Source: Authors' calculations based on the USPTO Cassis CD-ROM, December 2006 version.

terms of the degree to which they build on prior inventions and extend it, ECA patents created through multinational sponsorship are better connected to global R&D trends and generally represent inventions of higher quality. Science and innovation policy in the region should encourage ECA countries to promote international collaboration and support a greater role for the private sector in knowledge generation. However, measures to support R&D are ineffective when the key prerequisites—human capital and investment climate—are insufficient and less competitive.

On the second question, the answer is: "extremely important." Indeed, we have new evidence of the importance of trade openness, FDI, human capital, R&D, and knowledge flows for innovation and absorption in ECA countries. Data from surveys of ECA firms suggest that the purchase of foreign capital goods is a major source of acquisition of newer, more effective technologies, and that vertical FDI promotes learning by local firms. The relationship between training and skills on the one hand and successful technology absorption on the other is complex, with causality almost surely running in both directions. Further, contrary to many studies that suggest that a transition of firms into exporting is not associated with an increase in higher total factor productivity, we find exactly the opposite—as shown by increases in measured technology upgrading.

The Serbian case studies (at the end of chapter 2) further suggest that absorption requires tough decisions and large investments, as firms need to spend resources on modifying imported equipment and technologies, and reorganizing production lines and organizational structures. The case studies highlight the important role of foreign investors in knowledge absorption, whether acquired through capital goods imports, exporting, hiring consultants and other knowledge brokers, or licensing technology. They also show that exposure to foreign markets fosters learning—and that this learning effect is not limited to foreign-owned firms.

Thus, encouraging innovation will first require improving the investment climate for innovative firms, which includes reinforcing the regulatory reform agenda, removing barriers to competition, and fostering skills development. In parallel, policymakers should adopt policies to spur participation in world R&D, as collaboration with researchers and multinational corporations abroad is an effective way to tap into the global knowledge pool, enabling both technological and intellectual transfer of know-how. The policies include: a collaboration-friendly intellectual property rights regime, subsidized exchange study abroad for scientists and those with doctoral degrees, free immigration of researchers, and incentives for multinational corporations to establish their R&D centers in the host country.

Connecting research to firms

The pretransition silo system of industrial enterprises on one side and RDIs and Academies of Science on the other, with universities limited to teaching, has proven woefully inadequate. In the 1990s, all the transition economies went through a painful process of privatization and restructuring of the old state-owned enterprises. This process took place in parallel to building the investment climate for new companies, SMEs, and FDI. But little, if any, restructuring was done in the R&D sector. The delays in restructuring of state-owned RDIs will likely be blamed on scientists' concerns of losing national R&D funds and on an antireform lobby by RDI directors. During the 1990s, "oligarchs" bought centrally located RDIs with prime real estate in large cities only to establish shopping malls, and fire the researchers.

Recent economic trends have exacerbated matters. The transformation of industrial enterprises following privatization, downsizing of traditional sectors facing sharp competition from emerging economies in Asia, and rapid growth of the service sector has led to a widening disconnect of the private sector and an innovation infrastructure that remains mostly public and has not evolved to meet new demands.

The commercialization of R&D outputs entails the transformation of inventions into new products, processes, and services that are developed and brought to the market by startups, spinoffs from existing companies, or existing companies. The main driver of this process in advanced knowledge-based economies is the private sector, which has a better understanding and capacity to turn R&D outputs into applications that can be successfully marketed.

In many OECD countries, RDIs occupy an important role in the national innovation system. Originally established to provide support to large state-owned industrial enterprises, today RDIs' clients often include small firms that lack the capability and market intelligence to identify their own technological, organizational, and managerial needs. But supporting this market segment requires specific skills in marketing and business that many universities and research institutes are unlikely to have. Moreover, the SME market is typically very fragmented and rife with market failures, so even successful market-driven RDIs rely on government programs to support SME demand.

The reality in ECA, however, is that because firms do so little R&D, government R&D plays a larger role than in the OECD. In OECD countries, 63 percent of all R&D is funded by industry and 30 percent by government. In ECA countries, the proportions in financing are reversed: 30 percent industry and more than 60 percent government.

Support for commercialization should prioritize economic viability above other considerations and be a mechanism to change the incentives of RDIs and create partnerships with the demand side of innovation. Indeed, RDI restructuring is essential to *level* the private–public playing field and prevent RDIs from "crowding out" innovative startups and spinoffs. Public support for private innovation activities should be overseen by government ministries or departments that relate directly to companies and the business sector—usually the ministry of economy, or a department of industry or economic development—rather than the ministries or departments of science or education. Moreover, the business sector should take part in administering and selecting the programs.

For RDIs to effectively integrate their activities in line with private sector innovative efforts, they will need to tackle some major obstacles identified in our case studies—ranging from external factors (such as mechanisms to finance industry-oriented R&D being underdeveloped or a poor investment climate reducing the demand for private sector innovation) to internal factors (governance issues such as RDIs lacking the institutional autonomy to operate efficiently or the private sector not being involved in the RDI's strategic decision-making process). While some of these factors can be addressed through policy reforms, others will require significant capacity building (figure 4). And addressing most of these them will require overcoming important legacy challenges.

FIGURE 4
A call for policy reforms and capacity building

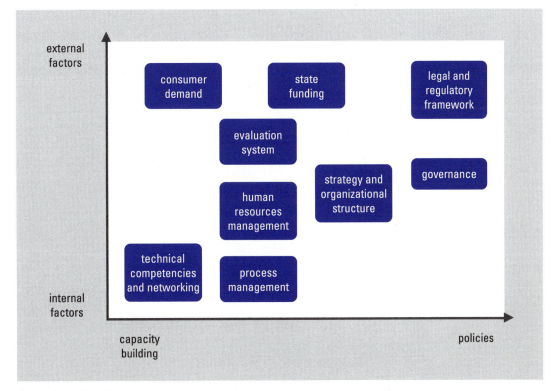

Source: Authors' analysis.

Restructuring options for RDIs

To connect research to firms, irrelevant RDIs need to be restructured, closed, or privatized. Our book—drawing on the World Bank's extensive experience in enterprise reform combined with an in-depth study of ECA and industrial country RDIs—suggests potential strategies for restoring the relationship between R&D and productive activities and tilting the public–private R&D balance:

1. Corporatization to increase autonomy but with government-ownership.

2. Insider restructuring while preserving government ownership.

3. Insider privatization.

4. Government-ownership but a change in management to contractor operated.

5. Transformation into a nonprofit foundation.

6. Outsider privatization.

7. Liquidation/closure.

Of these strategies, 1–3 are politically feasible but least effective in terms of governance reform; 6–7 are most difficult politically but sometimes urgently needed to reduce wasteful public spending; 4 and 5 have proved successful in the EU and the United States, but achieving the right governance and incentives may not be feasible in ECA because of corruption and regulatory capture concerns.

That said, the most realistic, and possibly novel, option for ECA policymakers may be the sale of shares to managers and researchers, so-called insider privatization. This option—with real estate ownership excluded because it will be subject to long-term land leases to avoid abuses—could be applied to complete the restructuring of commercialized RDIs and those volunteering for privatization.

Bringing innovations to market

Governments can support bringing innovation and technology to market in a variety of ways. A first step is for them to design instruments that promote private risk-taking and stimulate markets for private risk capital. Good instrument design principles include providing a good institutional environment (investment climate, intellectual property rights legislation, and avoiding corruption and regulatory capture in project selection by using international peer review); not crowding out private investment and other funding sources (by requiring the matching of public funds with private cash contribution—"additionality"); and minimizing distortions ("neutrality").

Government interventions should be designed to promote private risk taking and stimulate the private risk capital market. Four design issues should be taken into account:

- *Risk sharing.* The high uncertainty about technological and commercial success in the early phase not only deters mainstream financial institutions but it also represents a risk for the innovator. Often, the inherent uncertainty of success is the key obstacle in providing incentives to potential entrepreneurs to invest their own money, accommodate the opportunity costs of leaving a secure job, and take commercial risks by borrowing money.

- *Preservation of incentives.* The design of the instruments needs to preserve the incentives for entrepreneurs to invest their intellectual resources and time and effort in the pursuit of success. Concessionary funding is prone to "moral hazard" problems.

- *Commercial orientation.* Criteria for funding decisions need to clearly distinguish between projects that are technologically interesting and the targeted group of projects that are technologically innovative and have potential for commercial success. Commercial-success potential must be a criterion for project selection.

- *Specific bottlenecks.* The choice of instrument varies according to the different stages of the innovation chain. In some ECA countries, the most effective set of interventions will be combinations of financing instruments and measures to enhance innovative capacity and reforms to the business environment. The optimal level and degree of subsidy should be lower, the closer the intervention target is to functioning market mechanisms.

The government also needs to carefully sequence its provision of support for innovation—ensuring support is provided throughout the entire commercialization cycle, but with different policy instruments in the three main stages:

- *Early stage*: tapping incubators, angel investors, mini-grants or matching grants, as well as spinoffs and other spillovers from multinational corporations.

- *Growth stage*: providing government support for private venture capital through risk sharing.

- *Mature stage and exit*: facilitating access to international and local equity funds and strategic investors.

At the *early stage*, mini-grants and matching grants have been successfully used in developed countries—such as the U. S. Small Business Innovation Research Program, Finland's National Technology Agency of Finland, Israel's Office of the Chief Scientist, and Australia's Commonwealth Scientific and Industrial Research Organisation—while incubators have been used extensively, with mixed results, in the EU and the United States. But in ECA, attempts to adopt grants or incubators have been plagued by a lack of transparency and neutrality in project selection, a focus on short-run gains, and possibly corruption.

What can be done? It is crucial to strengthen the implementation arrangements of institutions supporting innovation (which include not only incubators but also technology transfer offices, and science, technology, and innovation parks), and enhance grant schemes by linking them to competitive funds that are peer reviewed by panels including international experts. Coordination failures—for example, when innovation efforts are duplicated in several institutions, or when R&D results are not accessible to knowledge users—can be overcome by innovation instruments that encourage "consortia" of universities or research institutes and firms both local and foreign.

At the *growth stage*, government programs to support venture capital have had mixed results in developed and Asian economies. A case in point is the failure of state-owned venture capital to attract a sufficiently large deal flow of projects, as in the initial stages of the Russian Venture Company in Russia. What can be done? To support the emergence of venture capital at this stage, rather than taking the counterproductive path of majority ownership, the government can mitigate private investors' risk by investing as a founding and passive limited partner, with minority ownership.

Effective sequencing should aim at building a significant pipeline of early stage projects before supporting venture capital. If not, the venture capital will likely fail as it will lack a sufficient number of projects to create a good portfolio. The success of the growth stage depends on a deal flow of attractive companies coming out of the early stage.

But setting up one or more programs may not be sufficient to trigger the entrepreneurial response. Consider Finland, which has also implemented extensive R&D support programs for more than three decades and is widely regarded as a prominent success in developing a thriving high-tech sector. Yet, a report on the Finnish innovation systems complains that there is still a noticeable shortage of innovative entrepreneurship (Georghiou and others 2003). Even more striking is Chile, where a

World Bank project focused the support of innovation on the promotion and funding of venture capital. But the well-funded venture capital program lacked a sufficient pipeline to invest in, and the project was not successful in raising R&D and innovation levels in the Chilean economy.

The difference between necessary and sufficient conditions is of utmost importance here: the establishment of a direct support program for R&D is in many cases necessary for innovation, but it is by no means sufficient. It can (also) have significant signaling value, showing serious commitment by the government to promote innovation and R&D over the long run and, hence, making it worthwhile for the entrepreneurs to engage in innovation activities. But it can also fail if the requisite institutional conditions are not in place.

CHAPTER 1

Why innovation matters—and what the government should do about it

- ▶ R&D spending in ECA countries remains low, standing at an average R&D-to-GDP ratio of 0.9 percent, about half of the EU27 average of 2 percent. The Czech Republic and Slovenia, the ECA economies with by far the highest R&D intensity, spend just about 1.5 percent of GDP.

- ▶ Innovation outputs are comparatively low in ECA, even when considering the level of R&D inputs, which reflect institutional and capacity weaknesses and the low share of private R&D. Even the leading ECA country in terms of granted patents in USPTO, the Russian Federation, lags behind OECD and other large middle-income countries in R&D output, including a much lower number of patents and scientific publications per capita.

▶ **Does government intervention have a role in raising public R&D and stimulating private R&D, as well as commercializing new ideas and absorbing technologies from around the world? The answer is yes, but a qualified yes: well-targeted government interventions can moderate several market failures affecting investment in innovation but can also aggravate them if badly planned and executed.**

▶ **Should support instruments be neutral about sectors? Although some OECD countries apply a mix of technology-neutral and technology-specific approaches, weaknesses in the governance and institutional framework in ECA countries often distort the allocation process, rendering them subject to enormous pressures from vested interests.**

As policymakers in Europe and Central Asia (ECA) debate ways to increase and maintain productivity and economic growth—and speed up convergence with Europe—they need to find ways to create an environment that is conducive to the application of knowledge in the economy through innovation and learning. The history of excellence in learning and basic research in several ECA countries provides some basis for hope that commercial innovation could be adopted and built "on the shoulders" of the past. Translating this research foundation into economically productive commercial applications, however, remains a critical missing link in ECA countries. Against that background, this book focuses on public policies for building supportive knowledge institutions and creating an incentives framework for the support of commercial innovation.

We distinguish between innovation and technology absorption as follows. The Organisation for Economic Co-operation and Development (OECD) Oslo Manual identifies four types of innovation: *product innovation, process innovation, marketing innovation,* and *organizational innovation.* These innovations can be new to the firm, new to the market, or new to the world. The advantage of using such a broad concept of innovation is that it includes all activities involved in the process of technological change. These range from identifying problems and generating new ideas and solutions, to implementing new solutions and diffusing new technologies. *Absorption,* a subset of innovation, is the application of existing technologies, processes, and products proved and tested in a new environment in which the processes have not yet been tested and the markets and commercial applications are not fully known (box 1.1). This distinction does not preclude important complementarities between innovation as a whole and absorptive capacity. Innovation promotes absorptive capacity because human capital generation and knowledge spillover effects associated with the innovative process build absorptive capacity. The ability of an economy to research and develop new technologies increases its ability to understand and apply existing technologies. Vice versa, the absorption of cutting-edge technology inspires new ideas and innovations.

BOX 1.1
Defining innovation and absorption

- *Innovation:* the development and commercialization of products and processes that are new to the firm, new to the market, or new to the world. The activities involved range from identifying problems and generating new ideas and solutions, to implementing new solutions and diffusing new technologies

- *Product innovation:* development of new products representing discrete improvements over existing ones.

- *Process innovation:* implementation of a new or significantly improved production or delivery method and implementation of a new organizational method in the firm's business practices, workplace organization, or external relations. This includes "soft innovation," such as reorganization of layouts, transport modes, management, and human resources.

- *Incremental innovation:* innovation that builds closely on technological antecedents and does not involve much technological improvement upon them.

- *Absorption:* the application of existing technologies, processes, and products proved and tested in a new environment in which the processes have not yet been tested and the markets and commercial applications are not fully known—that is, they are new to the firm. It is a subset of innovation.

- *Absorptive capacity:* a firm's capacity to assess the value of external knowledge and technology, and make necessary investments and organizational changes to absorb and apply this in its productive activities.

Source: Authors.

Yet, the adoption of existing technology through trade, foreign direct investment (FDI), or licensing is not guaranteed or cost free.[1] Firms and countries need to invest in developing "absorptive" or "national learning" capacity, which in turn is a function of spending on research and development (R&D). Thus, domestic R&D has a role in developing a firm's

---

1. Cohen and Levinthal 1989; Kinoshita 2000.

ability to identify, assimilate, and exploit knowledge from the environment—that is, enhancing the *absorptive capacity* of the economy.

Innovations may be undertaken by individual entrepreneurs or startup firms—with no existing market power—or by incumbent firms with market power. It is the new entrants or the firms with no existing market power that are popularly claimed to be more likely to undertake the most dramatic and revolutionary innovations. However, worldwide, most successful innovations are born, bred, and brought to market in larger incumbent firms with market power; often these innovations are incremental but nonetheless critical for sustained growth and job creation.

As for R&D, we use the widest definition to cover outcomes related to improvements in existing processes or products as well as the imitation and adoption of knowledge—it is not restricted to original innovation. The OECD defines R&D to "comprise creative work undertaken on a systemic basis in order to increase the stock of knowledge and the use of this stock of knowledge to devise new applications." Following the literature, R&D should be understood as "the process by which firms master and implement the design and production of goods and services that are new to them, irrespective of whether they are new to their competitors—domestic or foreign."

Our choice to focus on public support of commercial innovation is driven primarily by the increasing attention policymakers in the ECA region are directing toward enhancing investments in R&D in their respective countries—in other words, "client demand" for an analysis of the R&D commercialization support systems. The European Union's (EU) EU2020 (formerly Lisbon) Strategy prompted the EU accession countries and other ECA countries to consider implementing financial instruments to promote innovation, including venture capital schemes (in many cases, there was little consideration for the necessary institutional requisites or appropriateness of the instrument). In a number of countries in the former Soviet Union (for example, Russia, Ukraine, and Kazakhstan) and in other post-transition countries, the legacy of research and human capital also provides an incentive to revive their research capacity. However, absorptive capacity remains an issue in all ECA countries. Some of them are likely to have higher productivity returns from investments in building absorptive capacity than in commercial innovation.

The current allocation of research funding contributes to the apparent lack of collaboration between the science and business sectors. The aim of the financial instruments we recommend is to address those problems through the encouragement of private R&D in companies by providing incentives for collaboration through the cofunding of "consortia" of firms and universities or research institutes to implement innovative projects.

Nonfinancial instruments, such as business support services, incubators, and economic support zones, are suggested as complementary components of the financial instruments. A sound investment climate is considered an essential prerequisite for innovation and absorption.

The focus of this study on R&D and on commercialization is consistent with the view (elaborated in chapter 3) that commercial innovation and R&D are key factors driving *self-sustained*, long-term economic growth and, moreover, that these factors are generated from within the economic system, responding to economic incentives.

Although there is a wide variation in the intensity of R&D spending across countries, it tends to be lower in ECA countries—where the average R&D-to-GDP ratio is 0.9 percent (figure 1.1) and the world average is 2.4 percent. Some Nordic countries like Finland and Sweden spend about 4 percent of GDP on R&D, which is around double the OECD and EU27 averages. The large OECD countries including the France, Germany, the United Kingdom, and the United States have an R&D intensity of between 2 percent and 3 percent. The Czech Republic and Slovenia, the ECA economies with by far the highest R&D intensity, spend just about 1.5 percent of GDP. And human resources for R&D are unevenly distributed in ECA, though most ECA countries have a higher stock of researchers than other middle-income countries such as Brazil and Malaysia (figure 1.2).

Moreover, innovation outputs are comparatively low in ECA, even when considering the level of R&D inputs. Russia lags behind OECD and other large middle-income countries in R&D outputs—including a rela-

FIGURE 1.1
ECA needs to boost its R&D spending

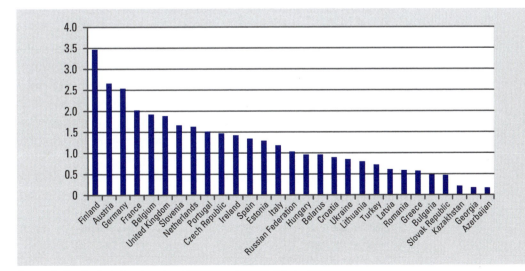

Source: Authors' calculations based on UNESCO and USPTO data for 2007.

FIGURE 1.2
ECA's researcher population is unevenly distributed

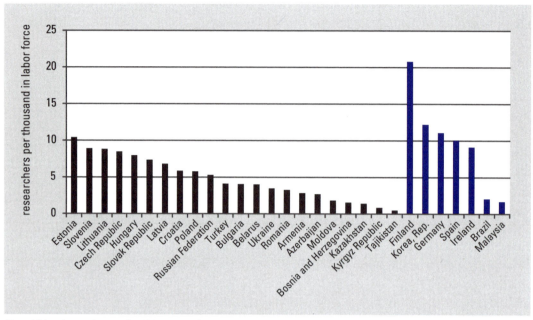

Note: All headcount data are for 2007, except for Malaysia, 2006.
Source: UNESCO statistics.

tively low number of patents and scientific publications per capita.[2] In fact, they spend more on R&D than most EU15 countries for each U.S. Patent and Trademark Office (USPTO) patent they generate (figure 1.3). Hence, ECA's low patenting performance can be attributed to both its low level of investments in R&D and inefficiencies within its innovation system. It must be emphasized that USPTO patents measure only innovation at the global technological frontier, while much of the R&D in ECA countries is used for catching up purposes. Radosevic (2005) does not find inefficiencies in R&D in ECA countries when considering national patents, but comparing national patents has its shortcomings since patents may be easier to obtain in one country than in another. However, in his empirical work, Radosevic does find inefficiencies in converting measures of innovation outputs, such as patents, into productivity in ECA countries. As a result, ECA countries are quickly being surpassed by China and India in terms of patenting (see figure 2.5, chapter 2).

Financial support to stimulate commercial investment in R&D by firms is important in ECA because the average R&D-to-GDP ratio does

2. Schaefer and Kuznetsov (2008) show that despite Russia devoting significant resources at the aggregate level to R&D, it is not translated into higher levels of total factor productivity. The authors suggest that the coexistence of a large R&D sphere and low productivity in manufacturing indicates low productivity in R&D institutions and weak links between R&D and the economy.

FIGURE 1.3
ECA's R&D efficiency is still low

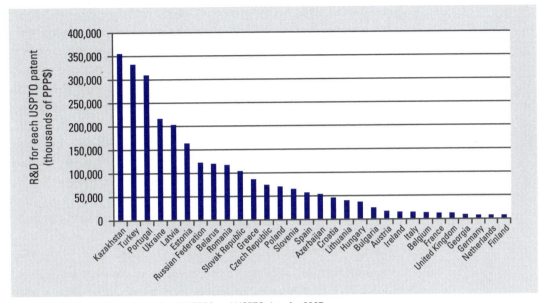

Source: Authors' calculations based on UNESCO and USPTO data for 2007.

not reveal the whole picture of the structural misallocation of resources between private and public sectors and between basic and commercial R&D in the transition economies. Typically, the bulk of R&D spending in ECA, as much as two-thirds of the 0.9 percent of GDP, is financed by governments, whereas only about one-third is financed by the private sector. By contrast, in countries with high rates of R&D expenditure, such as Finland, Germany, Japan, Ireland, Sweden, and the United States, the share of industry-related R&D spending ranges from 65 percent to 70 percent, whereas government spending amounts to only 20–30 percent (OECD 2002).

This chapter's goal is to provide countries with a general guide for evaluating the instruments to support innovation and the necessary institutional requisites for its effective application. Another key message is that ECA countries need to analyze the state of their national innovation systems before embarking in the adoption of many of the financial instruments pursued by EU countries to support innovation. Some countries may not meet the basic institutional requisites, such as economic incentives, education, and information infrastructure, to absorb innovation instruments effectively. Other countries may have institutional bottlenecks that need to be addressed before or concurrently with embarking on an innovation program.

The rationale for innovation

Ever since the path-breaking research of Robert Solow in 1956, economists have known that a country's long-term economic growth stems mostly from technological change, rather than traditional inputs such as capital and labor. Indeed, a vast array of subsequent empirical research over a half century has shown conclusively that at least half the growth in per capita income, in virtually every country studied, is associated with the growth of what is called total factor productivity (TFP)—that is, the famous "residual" to which we attach the label of technological change. But what exactly does this residual contain, how does it evolve over time, and what is the nature of the economic forces that determine its course and pace?

Indeed, one of the frustrating aspects of the early phase of economic thinking about these matters was that the growth of TFP—arguably the single most important economic phenomenon—appeared to economists as an impenetrable "black box" and seemed to occur outside the realm of economic forces. A long and fruitful research agenda pioneered by Griliches, Jorgenson, Denison, Rosenberg, and their associates sought to open this "black box" to understand its contents. However, it was only with the extensive development of endogenous growth theory in the late 1980s (Romer 1986, 1990; Lucas 1988; Grossman and Helpman 1991) that the economic profession came to accept the view that innovation, spillovers, and R&D were indeed the key factors driving self-sustained, long-term economic growth and, moreover, that these factors were generated from *within* the economic system, responding to economic incentives.[3]

In recent years, many studies have explored the interplay between competition and innovation, along with their impact of growth (Aghion and others 2005). Although Schumpeterian growth models predict that only firms with market power would have the resources and incentives to innovate, these studies find that in the more developed economies, among the incumbent firms closer to the "technology frontier," competition does encourage innovation. As for transition economies, a further study (Aghion, Carlin, and Schaffer 2002) shows that competitive pressures raise innovation in both new and incumbent firms, subject to hard-budget constraints for incumbent firms and the availability of financing for new firms. How about Europe as it edges closer to the world technology frontier? A recent study (Aghion and Howitt 2005) argues that Europe would benefit from a competition and labor market policy that

---

3. Work on education and technological change by Nelson and Phelps (1966) mentioned that technological progress was key to growth and highlighted the difference (for growth) between human capital stock and accumulation. However, it was only in the late 1980s that those views were widely shared.

not only emphasizes competition among incumbent firms but also stresses the importance of entry, exit, and mobility. The bottom line is that the closer firms in ECA countries move to the technology frontier, the more competitive pressures and market structures will play a role in the innovation capability of a country.

This conceptual framework molds our analysis: on the one hand, the view of the centrality of innovation and knowledge creation in the growth process, and on the other hand, the understanding that these are economic factors that may be shaped and influenced by properly designed economic policies. Building on that view, a recent study by the World Bank (Chen and Dahlman 2004) seeks to decompose "knowledge" into a wide array of indicators—each of which represents an aspect of knowledge—and assess their contribution to growth. The study, which covers 92 countries from 1960 to 2000, confirms that knowledge is a significant determinant of long-term economic growth. It finds that an increase of 20 percent in the average years of a population's schooling tends to increase the average annual economic growth by 0.15 percentage points. As for innovation, a 1 percent increase in the annual number of USPTO patents granted is associated with an increase of 0.9 percentage points in annual economic growth. And when the information and communication technology (ICT) infrastructure, as measured by the number of phones per 1,000 persons, is increased by 20 percent, annual economic growth tends to rise by 0.11 percentage points.

One corollary of the developments just sketched was the emergence of a soundly based and carefully articulated economic rationale for public support of R&D and innovation, which is by now widely accepted among academic economists and practitioners. The basic argument for public support of R&D is that innovation is a critical factor for growth (and hence, among others, for poverty alleviation), but a well-functioning market economy cannot generate by itself the optimal levels of R&D. There are two main sources of market failure with respect to R&D:[4] partial appropriability (owing to spillovers), which does not allow inventors to capture all the benefits of their invention, and information asymmetries—for example, the difference between the information that an

---

4. For a full list of rationales for state interventions in fostering knowledge creation, see the flagship study of the World Bank's Latin American and Caribbean Studies, de Ferranti and others 2003. It lists other important aspects of knowledge creation that prevent markets from generating the optimal level of knowledge: the long-term and risky nature of R&D investments, lumpiness of innovation, and coordination failures. See Baumol (2002) for a description of the features of the free market economy (market structure, productive entrepreneurship and rule of law, markets for technology trading, and reasons why R&D expenditure might be efficient despite substantial externalities on innovation) that explain its effectiveness in promoting innovation and growth.

inventor looking for financing has about an invention and the information that the potential financier has. These failures inhibit private firms from investing enough in innovation and R&D, thus depriving the economy of one of the key levers for sustained growth.[5]

Coping with spillovers

A basic feature of knowledge creation is that the returns from investments in it are not fully *appropriable* by the original investor. Knowledge has significant public good attributes—that is, once created it can be used repeatedly by multiple actors at no or very low extra costs. Firms making investments in knowledge creation capture only a portion of the benefits created. They do not receive compensation for the "spillovers" that their innovative efforts generate—that is, for the positive externalities of their actions on other firms and agents. Further, new technologies confer benefits to the purchasers of new products (consumers and producers alike) that often exceed any increase in the selling price that can be sustained. These nonappropriable benefits are also referred to as *spillovers* to consumers and are of particular importance in the context of "general purpose technologies." Both types of spillovers, namely, the purely technological externalities and the excess benefits to buyers, imply that the social returns from innovations may be far larger than the private returns (Jaffe 1998).

As a result of this gap, innovators operating in a market economy will invest in R&D less than the socially optimal amount; the extent of underinvestment depends on the extent to which social returns exceed private returns, and that may vary widely across fields, technologies, stages along the innovation cycle, and so on. Empirical studies have shown that typically the social rates of return of R&D expenditures are very large, often several times larger than private ones (Klette, Møen, and Griliches 2000). Moreover, these studies show that the returns from R&D exceed by a wide margin the returns from other types of investments, particularly from investment in physical capital. This implies that a government role in increasing the amount of resources devoted to R&D at the economywide level can have significant social benefits.

Spillovers may occur in many different ways, one being the mobility of R&D personnel and entrepreneurs. The process of innovation and its commercialization in an enterprise builds the human capital of its employees. Employees acquire R&D skills and an understanding of tech-

---

5. Clearly though, it is not enough to spell out such an economic rationale: for it to lead to policy, it must be weighed against the *costs* of government intervention, namely, the well-known problems of "industrial policy," capture, and corruption, which constitute the so-called government failures.

nologies and markets that are partly general—that is, they go beyond the specific knowledge embodied in the innovation and protected by intellectual property rights. Employees that move from one firm to another carry with them this human (or innovation) capital, which may benefit their new employers beyond the increment in wages that the mobile employees may receive. If mobility takes the form of migration, then the origin countries may be unwittingly "subsidizing" the destination countries through these spillovers. Thus the mobility of R&D personnel and entrepreneurs is an important transmission mechanism for spillovers, and hence a channel that should be closely monitored because it may have both positive and negative effects on any given country.[6]

Openness to trade and FDI increases the probability of receiving spillovers that originate elsewhere. As Coe and Helpman (1995) have shown, large economies tend to benefit the most from international spillover flows mediated by trade. Countries can increase their productivity by importing goods (especially capital equipment) from foreign, more advanced technologies (Coe, Helpman, and Hoffmaister 1997). Another potential source of technology spillover is FDI, though investors frequently "keep their knowledge at home" (Blomstrom and Kokko 1999). That is beginning to change (for example, R&D is moving to India), though the international principles still maintain control of the innovations through patents registered abroad.

The impact of FDI is indirect, through "spillover effects,"[7] owing to the presence of multinationals—first, because they create links with domestic firms and, second, because their presence spurs domestic producers to invest in new technology to compete with the foreign-owned firms. For example, in the Czech manufacturing sector during 1995–98, the indirect effect of R&D through the development of the absorptive capacity was found to be far more important than the direct effect of innovative R&D in increasing productivity growth of the firm; it was also found that R&D and intra-industry spillovers from FDI go hand-in-hand (Kinoshita 2000).

---

6. The spillovers-based argument clearly holds for large, mostly closed economies: being closed there is no risk of spillovers slipping out, and being large there is a high probability that at least some other local economic agents will benefit. In small open economies, spillovers may spill *out* of the country and benefit external firms and consumers rather than the local economy. Any policy designed to promote R&D should aim not only at increasing total R&D but also at increasing total R&D in a way that incentivizes *local* spillovers rather than external leakages, develops absorptive capacity, and ultimately affects the productivity of a wide range of sectors in the local economy.

7. Spillover effects (from neighboring countries or industries) arise when production affects the economic activity of other local firms or their employees. Positive spillover effects occur through the supply of new information, new technologies, managerial practices, and so on. Thus the "social" gain is larger than the profit or productivity gain made by the "source" company.

In Poland, so far, spillover effects leading to technology improvements in firms are observed only in a few industries, such as the auto industry, in which foreign R&D is high.[8] To be able to capture these international spillovers, the country needs to develop "absorptive capacity" (Cohen and Levinthal 1989), which entails, among others, investing in local indigenous education and innovation.

Another result of partial appropriability is "coordination failure." Often, technical advances in a given field require complementary advances by numerous distinct parties. Any one party may find that it is not worthwhile to develop one component of the system unless it can be sure that others will develop complementary components. If these parties do not have a mechanism to coordinate their investments, it is possible that no investment will occur. Government support may tip the balance such that multiple actors will invest in R&D independently. Innovation instruments can also be designed specifically to remedy coordination failures in innovation by encouraging "consortia" of universities/research institutes and firms or by promoting technology "clusters."

Coping with unequal information and the "funding gap"

A second source of market failure related to knowledge creation has to do with asymmetric information between inventors and external agents (for example, investors such as banks). Innovative activities entail by necessity a fundamental information asymmetry, certainly ex ante, that is, at the stage at which the inventor formulates the idea and seeks funds to develop it. It can be assumed that inventors have sufficient knowledge of the technology and of the details of the planned innovation, of their true abilities to carry it out, and of the efforts they are willing to put into developing the innovation. However, there will always be a significant gap between what the inventor knows and what an external agent can gauge, even if the information on those crucial matters is well documented.

In particular, there will be significant information asymmetries in that respect between the inventor and mainstream financial intermediaries, such as banks and institutional investors, who lack the capacity to verify the specific technical information and claims of the entrepreneur. Poten-

8. This is consistent with Kinoshita's (2000) finding concerning Czech enterprises' data—in oligopolistic sectors, such as electrical machinery and radio and television, there is a significant spillover rate as a result of having a large foreign presence. Also, R&D investment has a higher rate of return in these sectors. However, less oligopolistic sectors, such as food and nonmetallic minerals, show no evidence of spillovers despite the large presence of foreign investors in them.

tial investors will therefore be skeptical of the likely returns on investments in developing new technologies. Entrepreneurs who could offer attractive returns may have no credible way of conveying such potential to risk-averse investors.

The information asymmetry makes it hard for a creditor or equity investor to predict the returns from a potential investment in innovative ventures, which implies that such funding is not likely to be forthcoming. In the absence of demonstrated cash flows or other collateral, a typical startup company or individual innovative entrepreneur will not have access to traditional sources of finance—this is the so-called funding gap. At the most basic level, the funding gap implies that entrepreneurs face stiff constraints in the funding of innovations and thus will not invest (or will invest too little) in innovative projects that may have high social returns. This funding gap has been studied in most detail in the United States, but the findings have direct implications for the ECA region as well.

Early stage technological development (ESTD) is the most problematic phase in the innovation process and is defined as the link between invention and innovation, when a new product and market are identified. In this stage, product specifications appropriate to the identified market are demonstrated, and production processes begin to be developed, allowing estimates of production costs. At the end of this stage, the entrepreneurial venture has articulated a business case.

Figure 1.4 emphasizes the importance of internal financing by enterprises, government funding, and "angel investors" in the ESTD stage (Auerswald and Branscomb 2003). But most important, it emphasizes the virtual absence of more mainstream intermediaries such as banks, private equity, and other institutional investors. Although the percentages are for the United States only, figure 1.4 illustrates that, typically, even in one of the most advanced and innovative economies, early-stage finance of innovative projects is undertaken directly by firms, if they have the resources, or by very specialized institutions, with a significant role played by the government.

Not surprisingly, internal funds account for the biggest share of ESTD financing in the United States, because that is the most straightforward way of overcoming information asymmetries. Established enterprises know the track record of their own inventors/employees and, typically, have a better understanding of the market and the commercial potential of internally proposed innovations than do outside agents. Enterprises use the cash flows generated by established operations to finance innovation or source external funds on the basis of their balance sheet strength.

Angel investors are another important source of ESTD funding in the United States and to some extent in Europe. The term "angel investor" refers to successful entrepreneurs that look for new opportunities to

FIGURE 1.4
Corporate ventures and the government play a key role in the early stages

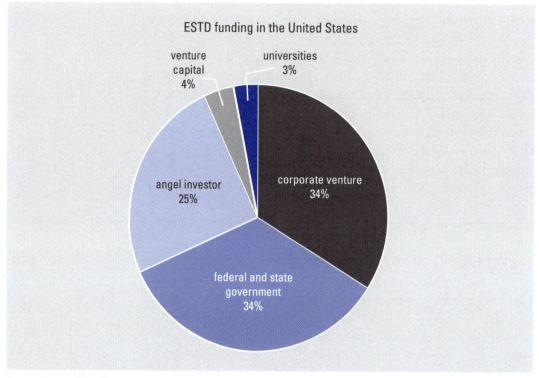

ESTD funding in the United States

- venture capital 4%
- universities 3%
- angel investor 25%
- corporate venture 34%
- federal and state government 34%

Source: Auerswald and Branscombe 2003.

invest private funds (earned from their own previous innovations) and are willing to invest in ESTD projects in technological fields that they understand well (having "been there and done that"). Studies of the behavior of angel investors frequently find that they are often heavily involved in the commercial decision-making and that this "business support" function can be as important as the financing. Managerial advice and commercial control over the ESTD entrepreneur are typical characteristics of the angel investor and venture capital funding models, as well as, of course, in internal funding models.

Given the short history of capitalist accumulation and profit-generating enterprises in ECA, internal financing by enterprises and angel investors is rare in the region and does not provide a viable basis for promoting innovation. The absence of angel investors is problematic not only from a funding perspective, but also given their role as sources of managerial expertise, as information brokers, and as access points to formal and informal networks of entrepreneurs and innovators. The role of government is therefore different in ECA countries than in OECD countries. The lack of "angels" and internal financing is acute, and the capacity of government agencies to fill their place is extremely limited. The Finnish case study in box 1.2 and the discussion in chapter 3 provide possible options for interventions that compensate for the absence of local angel investors by promoting international angel-investor networks and building public information marketplaces and networks.

ESTD requires patient and high-risk tolerant investment capital to fund early, prerevenue stages of research, development, and commercialization. Yet filling the "funding gap" requires specialized investors with the skills to evaluate and directly manage the risks of ESTD (angel investors or innovative managers in firms that are willing to invest retained earnings accumulated in other activities in the highly risky innovative projects) or governments with a broader public policy objective of capturing some of the spillovers associated with ESTD. In the absence of positive internal cash flows and angel investors, even if appropriability is adequate to yield a reasonable profit expectation, it may be impossible to secure the capital necessary to develop a new technology. Typically, in developing countries, the information asymmetry and "funding gap" problem is much more acute than in developed economies.

Why the government should play a role

In the presence of the markets failures discussed above, is there a case for government intervention in a market economy? The answer is yes, but a qualified yes: the necessary qualifications are about the *how* of government interventions.

The government plays a special role in promoting startups to generate new activities and support sustainable job creation because of the high risk that deters the entry of new ventures and the high failure rate once such ventures are established. This role derives from the asymmetry of risk between the government and the startup: for the private investor, the failure of a startup is a total loss, but from the society's point of view, that is not the case. In fact, the intellectual property assets that a failed startup created and the skills imparted to its former employees can be used to start a new enterprise utilizing those assets. Thus, for governments, the subsidization of failed startups contributes to innovation and the development of future startups. In particular, it is important that governments support new ventures that are based on intellectual property because those firms are, by definition, introducing new technologies and new products and developing new markets. Moreover, as Lerner (2009) points out, large firms often focus on existing clients, while new companies—faced with strong preexisting competition in established markets—often focus on developing and exploiting new market opportunities. Meanwhile, the dominance of these large firms in concentrated markets discourages the emergence of small innovative firms.

Thus, government interventions should promote the entry and growth of startups, particularly technology-based startups, through facilitating the commercialization of inventions and ensuring a level playing field between incumbent and small firms and the established large companies.

But while in a well-functioning market economy, there are institutions to facilitate effective government support and prevent abuses, in transition economies, government intervention might fail—or even cause harm—because of a weak institutional framework that is not conducive to intervention. As Josh Lerner of the Harvard Business School writes in a book[9] on entrepreneurship and venture capital: "When we look at the regions of the world that are, or are emerging as, the great hubs of entrepreneurial activity—places such as Silicon Valley, Singapore, Tel Aviv, Bangalore, and Guangdong and Zhejiang provinces—the stamp of the public sector is unmistakable. ... While the public sector is important in stimulating these activities, I will note that far more often than not, public programs have been failures. Many of these failures could have been avoided, however, if leaders had taken some relatively simple steps in designing and implementing their efforts" (pp. 5,7).

The bottom line is that any public intervention must be weighed against the actual and potential costs of intervention. Market failures may justify government intervention to stimulate absorptive capacity in the private sector. However, policy design needs to account for potential risks of government failures, such as corruption, the capture of policymakers by large companies and other vested interests, and misaligned incentives of government officials who risk high penalties if their policies fail but expect little extra compensation if they succeed.

"Industrial policy"

How deeply should governments become involved in picking winners and supporting the country's industrial champions? It is a timely question as governments are stepping up their use of industrial policy—that is, an attempt by the government to actively promote the growth of particular industrial sectors and companies—despite the controversy surrounding the topic. Critics point out that "picking winners" strategies have often failed, leaving taxpayers to foot the bill, or have been turned into programs that in reality are "saving losers" that could only stay afloat with subsidies.

One of the concerns is that successful program design often requires "neutrality," which is aimed at minimizing distortions. Neutrality means that the government does not "pick winners" and does not decide which

---

9. Lerner 2009.

sectors or technologies to support. On one hand, historical evidence finds that countries that made a successful transition from agrarian to modern advanced economies had governments that helped individual firms overcome coordination and externality problems. This holds true for the long-established industrial powers of Western Europe and North America as well as for the newly industrialized economies of East Asia (Lin and Monga 2010).

But industrial policy has also led to major and extremely costly failures in developing countries, with the government's attempt to pick winners and losers often the culprit. In the 1970s, for example, industrial policy was often associated with failed import substitution policies. A review of several recent major industry successes in developing countries by Pack and Saggi (2006) provides little evidence in favor of activist government policy. Take the cases of India's software sector, Bangladesh's clothing industry, and China's special economic zones. In the first two, the government's main role was one of "benign neglect," while in the latter China imitated the earlier success of Singapore by enabling the location of foreign investment in enclaves that were well provided with infrastructure. Much of the earlier investment came from overseas Chinese. In other words, these success stories were driven primarily by private sector agents (often from abroad). A further limitation on the potential role of industrial policies as traditionally understood, Pack and Saggi argue, is that many industries that developing countries would like to support are now highly globalized—making it much more difficult to set up and nurture national champions in isolation from existing international industrial networks and supply chains.

So does industrial policy have any role to play in economic development? Rodrik (2004) contends that the traditional view of industrial policy (based on technological and pecuniary externalities) is outdated and does not capture the complexities that characterize the process of industrialization. He argues that the right way of thinking about industrial policy is as a discovery process—one where firms and the government learn about underlying costs and opportunities and engage in strategic coordination. His view is that industrial policy is more about eliciting information from the private sector than addressing distortions by first-best instruments. He envisions industrial policy as a strategic collaboration between the private and public sectors—the primary goal of which is to determine areas in which a country has a comparative advantage. The traditional arguments against industrial policy lose much of their force when one views industrial policy in these terms. For example, the typical riposte about governments' inability to pick winners becomes irrelevant. The fundamental departure of this viewpoint from classical trade theory is that entrepreneurs may lack information about a country's comparative advantage. Or more to the point, at the micro level, entrepreneurs may simply not know what is profitable and what is not.

The relevance of Rodrik's argument (2008) hinges on the institutional innovations that can be put in place to cope with information asymmetry and rent-seeking.[10] Among the valuable ideas and principles for institutional development he proposes, it is worth highlighting "embeddedness"—which refers to the institutions that help governments work more closely with the private sector to discover the nature of market failures, so that final targeting decisions are truly guided by "a process of discovery." This means the use not only of carrots but also of sticks to weed out policies or projects that fail, and broader accountability of industrial policy to the general public.

But is this argument sound? Brahmbhatt (2007) has argued that there is a circularity problem in Rodrik's hypothesis that second-best policies, such as industrial policy, are needed to address market failures affecting modern sector activities because first-best policies like strengthening governance and building institutions are too broad and unrealistic. If it is true that to make industrial policy work there is a need for quite sophisticated governance and institutional mechanisms, then might not the original first-best policies also make sense? Perhaps only a few developing countries can muster the institutional strengths needed to make industrial policy work. At any rate, practical implementation would require close attention to the necessary governance and institutional underpinnings of industrial policy.

Lin and Monga (2010) argue that the discrepancy between the positive outcomes of industrial policy in advanced economies and its negative outcomes in developing countries lies in the poor choice of industries supported in the developing countries. Too often, instead of "picking winners," governments end up "picking losers." They argue that developing countries have tended to support industries that are too advanced and hence too far from the economy's comparative advantage (which might lie in labor supported capital-intensive industries). In contrast, emerging countries—that is, the rapid technology followers—have tended to support industries that were consistent with the comparative advantages in their economies, and typically similar to mature industries in countries whose income level closely paralleled their own.

So should innovation support measures to target firms in particular sectors? The answer needs to be nuanced. Two of the world's most advanced economic areas, the EU and the United States, apply a mix of technology-neutral and technology-specific approaches—with EU and U.S. enterprise R&D programs extensively engaged in "picking" technol-

---

10. Cited from Brahmbhatt 2007.

ogy areas.[11] Are these policies "picking winners" or, in Rodrik's terminology, state support for self-discovery in novel technologies? Countries sometimes make exceptions on neutrality for technologies thought to exploit a comparative advantage, or general purpose technologies thought to have particularly strong spillovers on the rest of the economy, such as ICT, and for technologies associated with public goods, such as health, food security, and climate change. In countries willing to facilitate firm self-discovery to promote diversification, support for innovation would need to be directed away from mature industrial sectors. This, of course, implies that governments have the capacity to identify those sectors with the highest positive spillovers. Moreover, in all cases, governments can rely on the fact that the private sector is willing to back a particular project as a minimal market test.

In some ECA countries with a closed economy or where there is no level playing field in the market, technological neutrality could be a mixed blessing. On one hand, it would ensure that governments do not pick overly ambitious sectors, and on the other hand, there is a risk that the enterprises, which are doing well because of a monopolistic position or subsidies, could submit innovative projects that seem profitable on paper due to the benefits from the above subsidies or monopoly. To avoid this distortion, project selection must be done from a social point of view by eliminating monopoly gains and subsidies from the project's anticipated cash flows.

Now that we have established the key reasons why innovation matters for boosting growth and living standards, the big question is how policymakers in ECA can improve the ability of their countries to tap into the global technology pool and how to leverage this knowledge to generate more innovations. As the chapter 2 explains, the process of knowledge absorption is neither automatic nor costless.

---

11. See http://cordis.europa.eu/themes/home_en.html#cloud and http://www.atp.nist.gov/atp/category.htm.

Acquiring technology from abroad—leveraging the resources of foreign investors and inventors

▶ ECA's inventive activity, as measured by patents, has been on the rise over the past decade, but it has been concentrated in just a few of the more advanced ECA economies. The relative isolation of ECA R&D is underscored by the relatively small number of citations these patents receive in patents subsequently granted.

▶ There is new evidence that foreign firms appear to be making a significant contribution to ECA inventive activity. Their local R&D operations and sponsorship of local inventors—known as international coinvention—generate a large fraction of the total patents emerging from the region.

▶ There is also new evidence of the importance of trade openness, FDI, human capital, R&D, and knowledge flows for innovation and absorption in ECA countries—reinforcing the need for a business friendly investment climate.

A key driver of economic growth, industrial development, and raising worker productivity is a country's absorptive capacity, or ability to tap into the world technology pool. But the process of knowledge absorption is neither automatic nor costless. Rather, it is often the case that extensive, active efforts are required to take technology pioneered outside the region and adapt it—in large and small ways—to the economic circumstances of Europe and Central Asian (ECA) countries.

Absorption is a costly learning activity that a firm can employ to integrate and commercialize knowledge and technology that is new to the firm but not new to the world. For simplicity, let us view development of new-to-the-world knowledge as "innovation." In other words, innovation shifts a notional technological frontier outward, while absorption moves the firm closer to the frontier. Examples of absorption include: adopting products and manufacturing processes developed elsewhere, upgrading old products and processes, licensing technology, improving organizational efficiency, and implementing a quality management system.

How does absorption occur? It requires dense links to the global knowledge economy, human capital, and a learning-by-doing process, among other factors. Trade, foreign direct investment (FDI), skills levels, human mobility, research and development (R&D), and flows of codified knowledge are "channels of absorption"—that is, they are central conduits for cutting-edge innovations. Figure 2.1 shows the important channels of absorption at the country level, as well as at the firm level.

Just as the prerequisite conditions and capacities are impacted by governmental policies, so too can properly designed economic policies influence the channels of absorption and the specific decisions for firms to absorb technology. Technology absorption needs a stable and conducive policy framework, and a firm's ability to absorb this technology and knowledge depends on its inherent characteristics, such as the vision of its CEO and owners, the quality of managers, the financial resources that can be directed toward R&D expenditures, and whether worker skills can adapt to changing production and marketing situations. Access to knowledge and technology is increasingly linked to FDI and trade. However, extracting benefits from these channels requires dynamic local firms and institutions. Absorption in the firm is thus determined by conditions internal to the firm, such as the presence of foreign investors, foreign trade, and skill endowments—and by conditions external to the firm, such as the costs and incentive structure determined by the wider environment, notably the regulatory framework; openness to knowledge flows, trade, and FDI policies; the quality, availability, and cost of infrastructure services; and the ease of access to finance.

This chapter takes as its primary focus the absorption of technology by ECA enterprises, rather than the creation of fundamental new technology. The reason for this is clear. Many ECA firms and industries lag well behind the global technological frontier, and relatively few ECA firms or industries are so technologically sophisticated that they could expect to play a leading role in the advancement of that frontier, at least in the near term. Given the level of development of the region's economies, it is almost certainly more important for managerial effort and public policy to be focused on convergence with the global frontier than on support of indigenous attempts at fundamental innovation. In some instances, the processes of modification and adaptation lead to innovations, often incremental in nature, dramatically increasing the value of the underlying technology in an ECA context. There are industries, firms, and regions within the ECA countries where the process of technology absorption has proceeded far enough that this kind of incremental innovation is taking place on a reasonably large scale.

FIGURE 2.1
Innovation and absorption spur growth and productivity

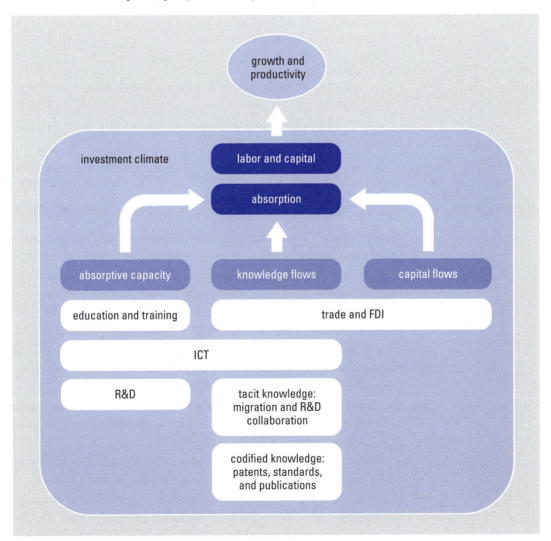

Source: Authors' adaptation from Goldberg and others 2007.

This chapter analyzes the extent of knowledge and technology absorption for firms in ECA, as well as the factors that influence absorption, using statistical analyses of various data sources, including the World Bank Enterprise Survey, the World Bank Investment climate Assessments, databases maintained by the U.S. and European patent offices, and case studies. The wealth of detailed information that patents and patent citations contain offers a useful window into the technological absorption process. While patents are indications of new-to-the-world innovation, much of this innovation is incremental, building closely on technical foundations developed in foreign countries. Patent citations connect ECA inventions to the prior foreign inventions upon which ECA inventions build, tracing pathways of international knowledge diffusion. These citations, included in patent documents, refer to "prior art" on which the invention draws from.

We ask two critical questions:

- To what extent ECA countries are able to leverage international knowledge flows and cross-national technological cooperation in the ECA region, as measured by patents and patent citations.

- The role of openness to trade, participation in global supply networks, and investment in human capital, through on-the-job training, to enhance knowledge and technological absorption in ECA-region manufacturing firms.

Cross-border knowledge flows

Scholarship on the most dynamic emerging economies tends to confirm that countries that are especially successful at closing the technological gaps vis-à-vis the high-income economies engage in at least incremental innovation and patenting. Economist Ryuhei Wakasugi's studies of Japan's imports of technology (Wakasugi 1986, 1990) noted the connection between Japanese firms' technology imports and R&D spending. Even in the 1960s, Japanese firms were generating impressively large numbers of patents at home and abroad (Evenson 1984). Trefler and Puga (2010) also point to the emergence of incremental innovation in contemporary developing countries such as China and present a model in which this incremental innovation, revealed by patenting, is a crucial part of the development process.

That patents should reflect innovation is obvious. That they may reflect certain kinds of knowledge absorption is less so, and it is important to explain why this is true. Many firms and industries in the region continue to lag behind the global technological frontier, and relatively few firms and industries in this region are currently playing a leading role in advancing that frontier. Given the level of development within the region's economies, it is almost certainly more important for managerial effort and public policy to be more focused on convergence to the best-practice frontier than on support of indigenous attempts at radical innovation. In some instances, the processes of modification and adaptation lead to innovations, often incremental in nature, that dramatically increase the value of the underlying technology in the context of a technology follower country. To a greater extent than is commonly realized, the major patent systems often grant patents that protect even relatively incremental innovations—both in terms of products and processes. These patents, and the wealth of detailed information they contain, offer a useful window into the ECA technological absorption process.

Almost by definition, successful patenting requires that a firm understand the existing state-of-the-art technology well enough to improve upon it, albeit perhaps in incremental ways. Simply by observing the firms, regions, and industries in which ECA inventors are most active, we can obtain objective, quantitative information on where the technological absorption process is most advanced. We can also observe how the locus of absorption and invention has shifted over time. The significant changes in ECA country patent regimes mean that the patent statistics of the region's countries themselves are unlikely to offer a consistent measure of inventive output over the course of the reform process. For that reason, we rely on data generated by ECA inventors seeking patent protection in the world's two largest patent jurisdictions: the European Patent Office (EPO) and the U.S. Patent and Trademark Office (USPTO). Both organizations provide large quantities of data on ECA-region inventors, obtained through a system whose essential features have remained stable throughout the transition period.

Data generated by USPTO provide a further unique window into the technology absorption process. Under U.S. patent law, all patent applicants are required to disclose knowledge of the "relevant prior art" on which they are built. These disclosures take the form of citations to earlier inventions and other technical advances that are often the technological antecedents of the invention for which the applicant is seeking patent protection. A large literature has utilized the citations in U.S. patent documents as direct indicators of knowledge spillovers.[12]

European patents also contain citations to prior inventions, but because European patent law does not require disclosure by the applicant, the vast majority of European patent citations are added ex post by

---

12. For example, Brahmbhatt and Hu (2010) use patenting in the United States as an index to assess East Asian prowess in generating innovations that advance the global frontier of knowledge.

patent examiners, and may or may not reflect inventions that were a source of inspiration to—or even known by—the actual inventor. Interestingly, detailed examination of the citation patterns in indigenous ECA patents reveals significant contrasts between ECA patents and those of other developing regions.

Indigenous ECA patents tend to be systematically less well connected to high-quality prior research than do patents from the more dynamic parts of the developing and developed world. This appears to validate the widely held view that ECA inventors, while highly skilled and well educated, are insufficiently connected to centers of technological excellence outside the region to reach their potential levels of research productivity.

General trends in ECA patenting

In the past few decades, the countries of Eastern Europe and the former Soviet Union have done much to close the significant gaps in industrial efficiency and technological sophistication vis-à-vis industrial Western Europe that developed during the decades of Europe's division. Even so, gaps still remain between these countries and the advanced industrial economies. To correctly prescribe policies for narrowing these gaps, we must first be able to identify and measure them across countries, indus-

FIGURE 2.2
ECA inventive activity on the rise

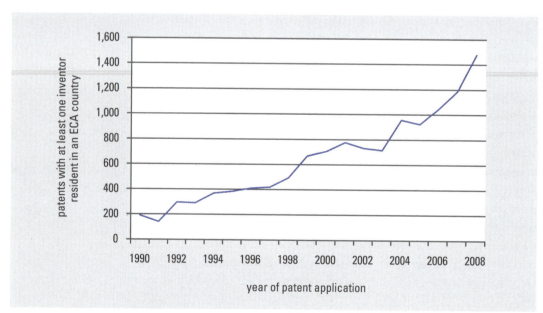

Note: Patent applications with at least one ECA inventor.
Source: Authors' calculations based on data provided by the USPTO, 2010.

tries, firms, and time. This analysis aims to show the importance of cross-country R&D collaboration and coinvention networks involving private and public partners as well as barriers that limit these exchanges and interactions between more and less technologically advanced countries.

First, it is clear that the transition process of the early to mid-1990s disrupted inventive activity in the short run. Both European and U.S. patents reveal a striking downturn in inventive output during these years. By the mid-1990s, however, measures of inventive output were once again trending upward, and this generally positive trend has been maintained until the most recent years for which data are available (figure 2.2).

Second, measures of inventive activity suggest a disproportionate concentration of that activity in the relatively more advanced ECA economies. Within ECA, there are five clear leaders: Hungary, the Czech Republic, the Russian Federation, Poland, and Ukraine. Among those, Hungary and the Czech Republic fare significantly better than the others (figure 2.3). Russia is a large patent generator in aggregate terms (figure 2.4) but less significant than one might expect given its size and scientific strength.

FIGURE 2.3
Hungary and the Czech Republic lead the ECA patents race

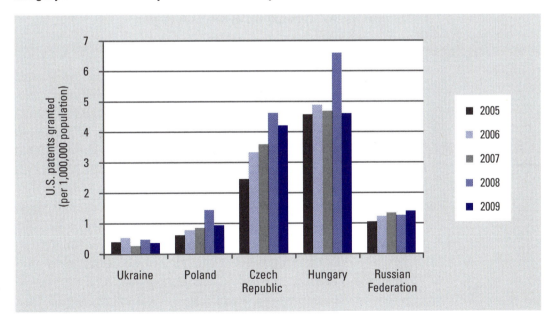

Note: The graph shows data for the five countries that have been granted the most patents from 2005–2009, with the Russian Federation as the leader in number of patents granted.
Sources: USPTO statistics and World Development Indicators, World Bank.

FIGURE 2.4

Russian Federation's patent share could be even bigger given its size and scientific strength

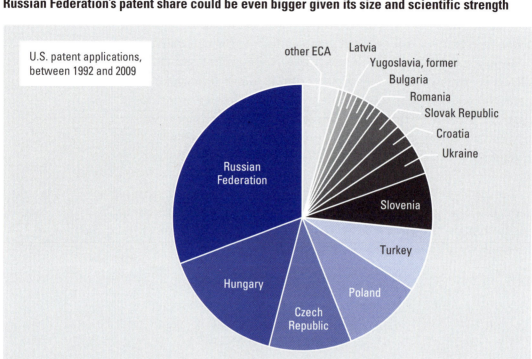

U.S. patent applications, between 1992 and 2009

Note: Patent applications with at least one ECA inventor, registered between 1992 and 2009.
Source: www.freepatentsonline.com, accessed August 8, 2010.

International coinvention in the ECA region

Foreign firms' local R&D operations, and their sponsorship of local inventors, collectively generate a large fraction of the total patents emerging from ECA countries. Not only does this process of international coinvention contribute to the quantity of ECA patents but it also raises the quality of ECA inventive effort. We designate a coinvention as a patent in which at least one named inventor is located in the ECA region and at least one inventor is located outside the region. Whereas indigenous ECA patents lag behind other regions in terms of the degree to which they build on prior invention and extend it, ECA patents created through multinational sponsorship are more effectively connected to global R&D trends and generally represent inventions of higher quality. Through collaboration with foreign scientists and engineers based in the world's innovation centers, ECA inventors are able to ground their efforts more solidly in the current technological state of the art and to benefit from knowledge of recent, relevant technical developments outside the region. The multinationals—unlike the state institutes—have both the incentives for and structure to promote the translation of the ECA research effort into formal intellectual property that can then be deployed both inside and outside the region.

Interestingly, the large role of international coinvention is a feature of regional inventive activity that the ECA countries share with India and, especially, with China. As with the production of goods, it appears that China, India, and ECA all have the opportunity to benefit from participation in an emerging international division of inventive labor, in which local inventors become part of a "production chain" of knowledge.

U.S. patent data confirm the importance of international coinvention as shown in the EPO data, both for the region as a whole and for individual countries. To place this in context, it is useful to compare recent invention trends for the ECA region with those of China and India given that the large ECA middle-income countries such as Russia and Ukraine—and to some extent Kazakhstan and Poland—are looking to China and India as benchmarks. Countries that have been through a transition—from central planning in China and a semi-socialist economy in India—are seen as interesting examples of overcoming the traditions of socialism. We show that just a few decades ago, the EU12 had the same number of patents as China and India, but this has been changing in recent years (figure 2.5). Currently, Russia holds roughly the same number of patents as all the EU12 countries combined.

If one sums up cumulative patent grants from 1993 to the end of 2009, the EU12 countries obtained 5,724 patent grants, whereas India-based inventors obtained only 4,759, and China-based inventors 6,817.

FIGURE 2.5
EU12 losing its edge on China and India

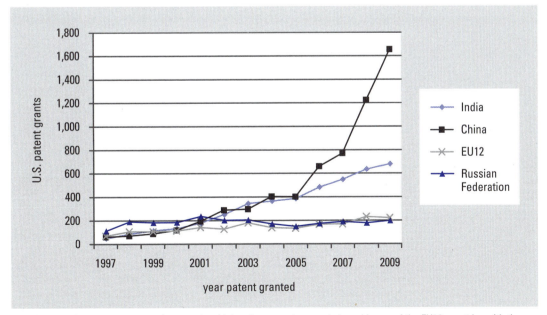

Note: The graph compares counts of patents in which at least one inventor is based in one of the EU12 countries with those of China, India, and the Russian Federation.
Source: Authors' calculations based on the USPTO.

FIGURE 2.6
The expanding role of international coinvention in the ECA 7

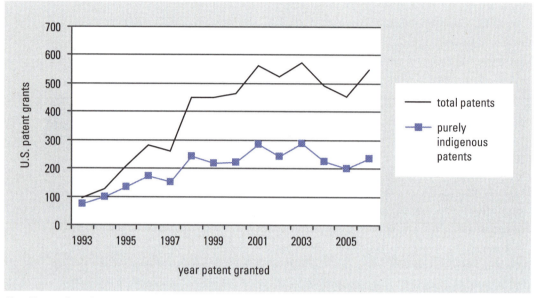

Note: The graph tracks total counts of patents in which at least one inventor is based in one of seven ECA countries: Bulgaria, the Czech Republic, Hungary, Poland, the Russian Federation, Slovenia, and Ukraine. "Purely indigenous patents" are those generated by a team whose members are all based in a single ECA country.
Source: Authors' calculations based on the USPTO Cassis CD-ROM, December 2006 version.

Clearly, when we normalize by population, the performance of the EU12 countries have been much better on a per capita basis. However, there is a clear acceleration in China-based and India-based patenting in the most recent years, which is not evident in ECA patenting.

How many of these ECA patent grants were generated by international teams of inventors versus solely through the efforts of inventors based within a particular ECA country? As figure 2.6 shows, international coinvention is *extremely* important—in recent years, more than half of total patent grants were generated from teams of inventors based in more than one country.[13] While we do see some inventive collaboration

---

13. Our data reveal a much greater degree of international coinvention than do earlier analyses. In part, this is because much earlier work has been based on the National Bureau of Economic Research Patent Citation Database, which ascribes the nationality of the patent to the location of the first inventor listed on the patent document. While incomplete, this method of assigning nationality generally "works" for patents generated in large, R&D-intensive economies like Japan or the United States. The overwhelming majority of Japanese and U.S. patents are produced by teams of inventors based entirely in Japan and the United States, respectively. But the usual method of assignment is much less appropriate for smaller, less R&D-intensive economies. The method misses the striking increase in the importance of international teams of inventors, including those based in these smaller economies.

FIGURE 2.7
Germany dominates ECA coinventions

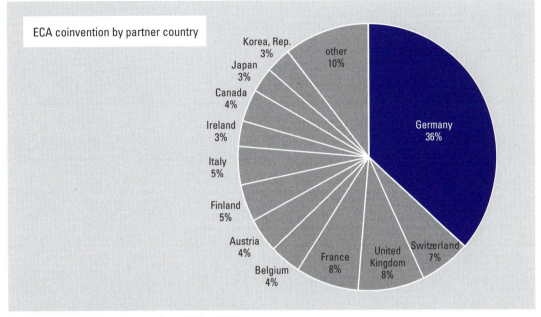

Note: Patent applications taken out from 1992 to 2005.
Source: Authors' calculations based on data from the European Patent Office.

between ECA countries, the patterns of coinvention are dominated by collaboration with inventors in more advanced countries. Germany plays a particularly important role (figure 2.7), followed by the United States, the other major European economies, Japan, and the Republic of Korea. We also see significant and growing international collaboration in the U.S. patent grants of China and India, but it is much less prevalent than in the ECA countries.

The ECA regional trend in international copatenting holds for its single largest member state. As figure 2.8 shows, most of Russia's U.S. patent grants are generated by cross-national inventor teams, and many are assigned to U.S. and other foreign multinationals. In fact, U.S.-based firms and other organizations generate about as many patents in Russia as do inventor teams composed solely of Russians. Table 2.1 lists the top 10 creators of Russia-based U.S. patents, with Japanese and U.S. multinationals figuring prominently. But this list is somewhat misleading. There are a large number of German firms that collectively account for a large number of patents, even though each individual firm has registered relatively few.

Are the coinvention results unique to the ECA region? As a basis for comparison, we carried out a similar study for Latin America and the

FIGURE 2.8
Expanding role of international coinventions in the Russian Federation

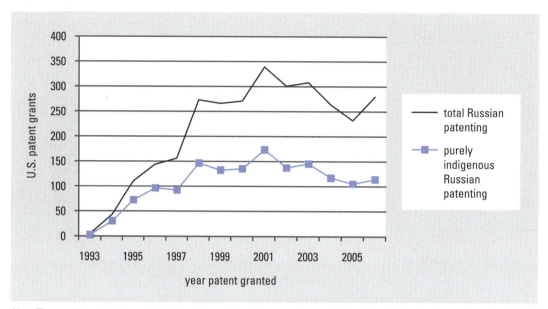

Note: The graph tracks total counts of patents in which at least one inventor is based in the Russian Federation. "Purely indigenous Russian patents" are those generated by a team whose members are all based in Russia.
Source: Authors' calculations based on the USPTO Cassis CD-ROM, December 2006 version.

Caribbean, which has numerous middle-income countries that are seeking to shift gears from resource-based growth to knowledge-based growth, and the results differ quite fundamentally along a number of dimensions. The four Latin American economies reviewed in detail (Argentina, Brazil, Chile, and Mexico) have all seen growth over time in invention and coinvention. In Argentina and Mexico, international R&D collaboration has become a quantitatively important component of overall patenting. However, the levels of all categories of patenting remain low, relative to Eastern Europe and Russia, and this is true both in absolute and per capita terms. The difference in industry mix is clearly an important one to keep in mind. Many Latin American countries are commodity exporters, and the agricultural and resource extraction sectors continue to occupy much larger fractions of GDP and employment than they do in most of Eastern Europe. This inevitably affects total patenting and patenting per capita, because agriculture tends to be a much less patent-intensive sector than manufacturing and many resource-extraction industries do not generate large numbers of patents. Additionally, Latin America does not benefit from the legacy of serious planning and colossal investments in science and technology that the USSR embarked upon in its effort to catch up and overtake the other major superpowers.

TABLE 2.1
Top generators of Russia-based U.S. patents

Rank	Name	Nationality	Number of U.S. patents
1	LSI Corp.	United States	111
2	Samsung Electronics	Republic of Korea	71
3	General Electric Co.	United States	37
4	Elbrus International	Russian Federation	36
4	Sun Microsystems	United States	36
5	Ceram Optec Industries	Germany	28
6	Nippon Mektron	Japan	26
7	Ajinomoto Co.	Japan	25
8	Procter & Gamble	United States	19
8	Ramtech	United States	19
9	Advanced Renal Technologies	United States	16
9	Corning Inc.	United States	16
9	Nortel Networks	Canada/United States	16

Note: The table tracks total counts of patents in which at least one inventor is based in the Russian Federation.
Source: Authors' calculations based on the USPTO Cassis CD-ROM, December 2006 version.

Therefore, its absorptive capacity and inventive performance have traditionally been lower than in ECA, both in the public and the private sector.

The high prevalence of international R&D collaboration evident in ECA patenting is likely to be a positive influence on the extent and nature of national inventive activity. Through collaboration with inventors based in more advanced industrial economies, ECA inventors are likely to achieve a greater understanding of recent technological trends and developments than would be possible through autarkic R&D effort. Patents that result from international coinvention with high-income economy partners tend to be of significantly higher quality than purely indigenous patents. Furthermore, there is evidence that the patents resulting from international coinvention are better, in part, because they emerge from an invention process that is more aware of, and integrated into, important recent technological developments outside of Eastern Europe.

What is driving the rise in coinvention?

Inventors in the EU8 seem to be increasingly taking part in an international division of R&D labor, one in which their own knowledge and skill are matched with the complementary intellectual and financial assets in high-income economies.

There are two key factors driving this increase. One is technological: prior to the Internet revolution of the 1990s, no economy possessed a communications technology infrastructure sufficiently powerful to allow parties on different continents to collaborate on something as intricate and complex as the creation of new technology. This technology is now globally available, and it allows the formation of international inventive teams.[14] Second, a positive legacy of socialism in Eastern Europe was a reasonably well-developed scientific and engineering educational system that continued to generate fairly large numbers of reasonably well-trained graduates into the post-transition economies. The reasonably high level of human capital combined with relatively low wages for the highly skilled has made Eastern Europe a relatively attractive place to source certain kinds of engineering and research tasks. In China and India, massive expansions of higher education systems in recent years combined with significantly lower wages have made these two Asian giants attractive alternatives sites for international coinvention.[15]

Is this rise a good thing for the host countries? It is conceivable that multinational firms could "buy up" local engineering talent and then employ the best local inventors in programs of research that benefit multinationals but not the host economies. This view would cast coinvention as akin to a "brain-drain" phenomenon—the engineers never actually leave their home countries, but the home economy is effectively deprived of their talent, while the multinational employer gets highly skilled professional labor at cut-rate prices. A more favorable view of the coinvention phenomenon would cast it as a way for domestic research workers to overcome the limitations of national innovation systems by integrating into a global system of innovation that connects Eastern European talent to foreign financial and scientific resources. Multinational partnerships raise the quality of ECA invention because ECA inventors working for multinational corporations are able to better tap external pools of relevant knowledge. Eastern European researchers and economies—such as Poland—are able to contribute to realizing Europe's research potential to

---

14. Interviews with Eastern European research professionals underscored the ease with which international coinvention can take place, thanks to modern communications technologies and low-cost international travel.
15. See Freeman (2006) for documentation of this stunning expansion of the educational system.

a much greater extent and at a much earlier stage than would be possible in the absence of international coinvention (box 2.1).

BOX 2.1
A snapshot of coinvention in Poland

To understand the interaction that lies behind the coinvention process, European Patent Office patent data were used to identify and interview the Polish members of international coinvention teams. While the numbers of interviewees were small, several themes emerged that provide insights into the process behind coinvention in Poland.

The genesis of the coinvention relationship. In some cases, it stemmed from employment with a multinational enterprise. Other times, the international scientific community played a big role. The participation of Polish researchers in international conferences and publication in international scientific journals often brought individual researchers to the attention of potential Western collaborators. The "diaspora" of Polish academics in the Western world helped greatly, as did extended periods spent abroad as postdocs or visiting researchers.

How each side benefited. A big draw for the Western partners was the relatively low cost of high-quality research labor in Poland. For Poland, a key benefit was that the Western partner often provided access to financial support for research activities and staff training beyond what would have been available from domestic sources. In some cases, the Western partner also provided useful expertise, though it was not always purely scientific—including an understanding of commercial applications and the importance of intellectual property. There were positive technological spillovers from collaboration with multinationals that outweighed any negative effects. And the possibility of collaboration with multinationals kept some researchers in Poland who would have otherwise emigrated—the remuneration and resources provided by multinationals permitted a level of research excellence that would have been difficult or impossible to achieve otherwise. In some cases, the collaboration process involved the multinational acquisition of domestic enterprises under terms that allowed the research activity to expand and improve.

Both sides actively participated. Most of the scientific or engineering work was conducted in Poland by Polish researchers, who often contributed the essential ideas and even played a leadership role in the project's engineering dimension. In other words, the coinvention process was not a case of outsourcing the relatively labor-intensive parts of the research to Eastern Europe. As for the Western partners, legal considerations alone suggest some intellectual input from them. According to European Union (EU) and U.S. patent law and practice, a firm incurs legal risks by omitting from a patent application inventors who made a significant contribution to an invention, or including on a patent application inventors who played no significant role.

Several problems emerged. First, Western partners sometimes sought sole ownership of the intellectual property emanating from the partnership and retained it under terms that provided for no royalties or other ongoing financial compensation to the Polish partner. Second, the high cost and expense of applying for patents and maintaining them in major international jurisdictions like the United States or the EU was well beyond the budget constraint of an individual inventor or a small firm. Third, public support in Poland for inventors and small firms to patent their inventions in international markets fell short of what was desired.

Source: Authors.

ECA patent citations

While ECA's coinvention activities point to global knowledge flows within the region, patent citation activity provides a different story. Examination of the citations in indigenous patents from ECA countries taken out with the USPTO reveals some striking contrasts with those from other technologically successful regions. In many cases, indigenous ECA patents make *fewer* patent citations, and the differences are often quite significant at traditional levels of statistical significance. While the presence of a large number of citations is sometimes viewed as evidence that the citing patent is less innovative, we believe that the low levels of ECA citations are indicative of an insufficient grasp of global best practices and recent advances in the state of the art. This interpretation is strengthened by the existence of another empirical regularity: indigenous ECA patents typically cite older patents than do patents generated in other regions. It is hard to explain this on the basis of the age of the patent or technology class, but we started with a sample essentially matched in both aspects. We also see that indigenous ECA patents tend to cite patents with a less broad technological impact than those in other countries.

All of this is suggestive of the notion that ECA inventors are insufficiently grounded in the recent technical state of the art. They are not sufficiently well versed in recent technical developments outside of their region to build on those developments with maximum effectiveness. This is further suggested by the fact that indigenous ECA patents tend to cite patents with a less broad technological impact as well as those that are older patents than do patents generated in other regions. Case-based and interview-based studies of R&D and productivity in the region have long criticized the tendency for centers of ECA scientific activity to be insufficiently connected to centers of technical excellence outside the region. Our data analyses, drawing upon data from thousands of individual patent grants from across the region, document citation patterns consistent with these criticisms.

Acquiring foreign technology

So how is knowledge absorbed? We know that openness to foreign trade and investment is critical to the process of technological absorption and diffusion not only for the competitive pressure that it exerts on management and corporate governance but also for the exposure to global best-practice technology and management techniques that such openness provides to local firms. The contribution of international openness to growth has been evident since the first wave of globalization in 1870–

1913. In his book, *The Mystery of Economic Growth*, Helpman (2004) focuses on four themes: accumulation of physical and human capital; productivity growth, as determined particularly by the incentives for knowledge creation, R&D, and learning-by-doing; knowledge flows across national borders and the impact of foreign trade and investment on incentives to innovate, imitate, and use new technologies; and institutions that affect incentives to accumulate and innovate. More recently, El-Erian and Spence (2008) cite "leveraging the global economy to accelerate growth" as one of the ingredients in the recipe that has the most potential to boost welfare and living standards.

To investigate whether *firms* benefit from such openness through exporting and inward FDI, we address two questions: Is there "learning-by-exporting?" and how does FDI affect absorption? Our interest centers on the impact of international trade, participation in global supply networks, and FDI as mechanisms to promote the diffusion and absorption of technology. Here, we need to draw a distinction between this phenomenon and the narrower concept of productivity "spillovers"—which is a benefit for which the recipient did not have to pay. Previous researchers have often presumed that the technology absorption abetted by trade or FDI would show up in the form of faster growth rates of total factor productivity: Through contact with superior foreign technology, indigenous firms would be able to realize large gains in output that are far greater than their investments in new capital or better workers.

Our broader concept of technology absorption or technology diffusion accepts the reality that indigenous firms may have to make costly investments to acquire new technical competencies, and their ability to appropriate the gains generated by these new competencies may be limited. In other words, a socially beneficial acquisition of new technological capabilities may take place, even when the conventionally defined "productivity gains" are limited or slow in coming, and it may take place even when the conventional techniques designed to measure "productivity spillovers" suggest limited effects.

How trade diffuses knowledge

Imports of manufactured goods, especially from more advanced economies, have been a channel of knowledge flows. By purchasing imported inputs and capital goods, firms in less developed countries acquire use of the technology embodied in these goods, thereby realizing gains in national welfare. Empirical multicountry studies[16] show that international trade mediates flows of knowledge, allowing firms and industries

---

16. Coe and Helpman 1995; Keller 2002.

to acquire technology that expands their productive capabilities, often in ways that show up in conventional productivity measures. In addition, a recent study (Acharya and Keller 2008), using a sample of industrialized countries over 1973–2002, shows that new competition from foreign firms raises net domestic productivity. This occurs because import competition triggers technological learning, along with market share reallocations between advanced imports involving foreign technologies and domestic firms with weaker technological capabilities.

This finding corroborates the long-standing belief among experts that trading with countries that have a richer R&D stock or, more broadly, that are able to export more advanced technology goods can facilitate the acquisition of new technical competencies on the part of importers. Conversely, one may posit that participation in export markets enables firms to become more productive, a phenomenon referred to as "learning through exporting," though this belief has not always been supported by empirical research at the firm level. In a similar vein, it is widely believed that trading in parts and components with foreign companies that are already well integrated in the global production network can facilitate the acquisition of new technology. Data from surveys of ECA firms provide direct evidence suggesting that the purchase of foreign capital goods is a major source of acquisition of newer, more effective technologies.[17] And historical accounts of the rise of East Asian export industries stress the role of advanced country buyers as conduits of technology and managerial know-how to developing country firms.[18]

Since trade and FDI are the key channels for international diffusion of knowledge, poor logistics may impede access to new technology and know-how, and in turn slow the rate of productivity growth. Logistics can be improved in a variety of ways. A study by Engman (2005) shows that improved and simplified customs procedures significantly help trade flows, and the facilitation of cross-border movement of goods helps a country attract FDI and better integrate into international production supply chains. Wilson, Mann, and Otsuki (2003) demonstrate that enhanced port efficiency and a removal of regulatory barriers sharply boost trade flows in the Asia–Pacific region—even more than better customs and greater use of e-business. Persson (2008), using data from the World Bank's Doing Business Database on the time required to export or import as indicators of cross-border transaction costs, finds timing is everything—on average, time delays on the part of the exporter and importer significantly decrease trade flows. Further, technical barriers to

---

17. In one question in the Business Environment and Enterprise Performance Survey dataset described below, ECA firms are asked to identify a single source (presumably the most important) of acquisition of new technology. The overwhelming majority cite the purchase of new capital equipment as the source of this new technology.

18. Pack and Westphal 1986.

trade significantly reduce the entry of firms into export markets, substantially reduce trade in some cases, and increase firm startup and production costs. And Chen and others (2006) shows that testing procedures and lengthy inspection procedures significantly reduce exports and impede exporters' market entry.

How FDI facilitates absorption

The close connection between FDI and trade is vital—a point emphasized in ECA's flagship trade study *From Disintegration to Reintegration: Eastern Europe and Former Soviet Union in International Trade.*[19] The benefits of international exposure to best-practice technologies often come directly or indirectly through the intermediation of foreign firms. The strength of this connection arises out of well-noted revolutionary changes in logistics, information technology, and manufacturing in recent decades, which allow firms to disaggregate the total production process of a good or service into separate stages, and to locate each of these stages in an environment where local factor endowments enable efficient production. This process, referred to as the slicing up of the value chain, or the fragmentation of production,[20] enables a broad range of countries to participate in various stages of the production chain—even if they do not possess intangible assets or a service infrastructure sufficient to provide them with a comparative advantage in the production of final goods. Multinationals have played an important role in driving this process—estimates suggest that about two-thirds of world trade by the latter half of the 1990s involved multinational corporations.

Multinationals have played particularly significant roles in expanding the international trade in some of the most successful developing countries. China clearly stands out over the course of the past decade, in terms of the speed with which its exports of manufactured goods have expanded and diversified across a broad range of product categories. China is now the second largest trading economy, as measured by conventional trade statistics (as of 2010), and it has recently emerged as the largest single supplier of information-technology hardware to the U.S. market. Most of the expansion of Chinese trade in the 1990s has been driven by foreign firms, and this is especially true of "high-tech" production. Today, foreign firms are responsible for nearly 60 percent of China's total exports. By 2003, foreign firms accounted for 92 percent of China's export of computers, computer components, and peripheral devices.[21]

---

19. World Bank 2005a.
20. Krugman 2000; Jones, Kierzkowski, and Lurong 2005.
21. See Branstetter and Lardy (2008) for an overview of Chinese trade and FDI reform, which stresses the role of foreign firms in China's export dynamism.

China is not alone in Asia in terms of its reliance on foreign firms to mediate exports of technologically intensive goods. In fact, China can be viewed as merely the latest manifestation of a phenomenon that is common throughout Southeast Asia. Athukorala (2006) presents an interesting view of Asian industrial history, in which he contrasts the industrialization of Japan and the East Asian newly industrialized economies, which were primarily based on the growth of indigenous firms, with the experience of "latecomer countries"—the Southeast Asian nations and China. He argues that a long period of import substitution in the latecomer countries undermined the development of indigenous entrepreneurship and cut off local managerial elites from international markets. It was simply more efficient for the latecomer countries to join a preexisting international production chain, rather than try to create one from scratch through the efforts of indigenous firms alone. Using official data from Asian countries to document the sizable expansion in the share of international trade mediated by multinationals in all of these countries, he concludes that the entry of multinationals is virtually essential for the export success of latecomers.

There is an ongoing debate in the literature about the impact of FDI: Some have suggested that positive technology spillovers from FDI are largely limited to "vertical" FDI transactions, in which there is a direct purchasing relationship between the foreign firm and the local supplier. Using direct proxies of the firm's relationships with multinationals, we find evidence that vertical FDI promotes learning by local firms—and our case study on Serbia identifies some explicit channels through which learning occurs. In addition, a recent study by Gorodnichenko, Svejnar, and Terrell (2010), using data for 27 transition economies, finds evidence that firms that supply a large share of their sales to multinationals tend to innovate more than firms that cater to the domestic market. Thus, ECA governments seeking to encourage technology absorption should continue to open themselves to FDI and should critically examine informal barriers to foreign firm expansion that might impede this channel of technology diffusion.

Reduction of the remaining barriers to FDI in ECA could increase FDI and, given the positive association of absorption and FDI, facilitate absorption. For example, Russia fares worse than other countries in the region, attracting one of the lowest levels of FDI inflows. Related World Bank-supported research has pointed to key shortcomings in the Russian business environment. Many of these shortcomings are a function of government policies that limit FDI inflows and foreign firm operations, especially in the service sector.[22] The reduction in barriers to FDI in ser-

---

22. Desai and Goldberg 2008.

vice sectors would allow all multinationals to obtain greater post-tax benefits on their investments, encouraging them to increase FDI to supply the Russian market. This, in turn, would lead to an increase in technology absorption, as implied by the positive association of FDI and absorption.

Removing barriers to trade in services in a particular sector is likely to lead to lower prices, improved quality, and greater variety. Efficient services would be vital intermediate inputs into the productive sector, and the telecom sector would be particularly crucial to the diffusion of knowledge. Technology transfer accompanying this service liberalization— either embodied in FDI, or disembodied, would have a stronger effect on growth.[23] Many investment climate studies such as Desai and Goldberg (2008) provide substantial and robust evidence that various measures of regulation in the product market, particularly entry barriers, are negatively related to investment. The implications of our analyses are clear: regulatory reforms, especially those that liberalize entry, are very likely to spur investment.[24]

Although technology is making it easier to trade in services, often FDI plays a vital role in selling services.[25] Given the lack of a service sector under central planning, FDI can be expected to play a particularly important role, more so than in countries where incumbent competition confronts foreign providers. Overall, services account for some 62 percent of the stock of FDI in 12 selected ECA countries, with finance, transport, communications, and distribution services accounting for the largest share of this FDI. While the share of the service sector in GDP, employment, output per worker, trade, and FDI in Central and Eastern European countries shows substantial convergence toward that of Western European countries, it also shows a distinct difference between Central European/Baltic states and Central Asian and Commonwealth of Independent States economies. Reforms in policies regarding financial and infrastructure services, including telecommunications, power, and transport, are highly correlated with inward FDI.[26]

---

23. Mattoo (2005) argues that since many services are inputs into production, the inefficient supply of such services acts as a tax on production, and prevents the realization of significant gains in productivity. As countries reduce tariffs and other barriers to trade, effective rates of protection for manufacturing industries may become negative if they continue to be confronted with input prices that are higher than they would be if services markets were competitive, making it imperative to have policies to liberalize trade in services and attract FDI in key service sectors like telecommunication and financial services.
24. Rutherford, Tarr, and Shepotylo 2007; Jensen, Rutherford, and Tarr 2007. Also see Alesina and others 2005.
25. Eschenbach and Hoekman 2006.
26. Eschenbach and Hoekman 2006, table 3.

Firm-level investments in human capital and the skill level of the workforce are strongly associated with technology absorption. In every regression, the presence of a worker training program was strongly associated with technology absorption, and measures of the skill level of the labor force often had significant effects on absorption. In the panel regressions, the introduction of a training program is positively associated with increases in technology absorption. The Serbia case study also highlights the importance of worker training in cases of successful technology absorption.

How worker skills training aids absorption

Human capital investment through training and higher qualifications is an important step toward achieving higher and longer term productivity at the firm level. These returns to the firms arise in two particular ways, namely internally and externally. Internal gains accrue to workers from skill acquisition but firms also gain to potentially gain due to increased labor productivity. External gains accrue to the firms in terms of spillovers from its interactions with skilled labor in the same local area.

The relationship between training and skills on the one hand, and successful technology absorption on the other, is complex, with causality almost surely running in both directions. Training and knowledge absorption are complementary, in the sense that a firm's capacity to absorb new knowledge, and to benefit from absorption, depends on the skills and training of the workforce. Higher levels of training and skills typically lead to a firm identifying new technologies that need to be mastered in order to increase competitiveness. Yet the decision of the firm to acquire a certain technical competency often necessitates training and changes in the skill composition of the workforce. For example, training in Russian enterprises is also highly correlated with indicators of innovativeness—such as R&D or licensing of patents and know-how, introduction of new production technologies, and high technology exports.[27]

Acemoglu and Ziliboti (2001) have highlighted the role of complementarities between human capital and technological progress. The authors suggest that differences in the supply of skills create a mismatch between the requirements of a given technology and the skills of workers. Thus even when all countries have equal access to new technologies, this technology–skill mismatch can lead to sizable differences in total factor productivity (TFP) and output per worker.

These findings have implications for both firm strategy and public policy in ECA. While all countries struggle to align the output of their formal public educational systems with the changing needs of their

27. Tan, Gimpelson, and Savchenko 2008.

industries, the challenge has been particularly acute in ECA. The legacy of socialism included a number of significant educational achievements, but many features of the prereform system were not well suited to the needs of an open, competitive economy. Despite the substantial progress that has occurred since transition, more work remains to be done. Again, in Russia, despite the high and rising demand for educated and skilled workers, there exist skill shortages in enterprises. The reasons for this shortage include an educational and training system that is underfunded below the tertiary level and that faces numerous challenges, an industrial sector with high labor turnover (which inhibits training), and the inability of some noncompetitive firms to pay competitive wages to attract and retain needed skills.

The issue of worker training also deserves consideration. In an economic environment with labor mobility, firms may be reluctant to invest in the skills of workers who might simply leave the firm and take those skills to a rival for slightly more pay. This is especially the case when firms face financial or other constraints that may limit their ability to engage in other necessary investments. While full consideration of this issue would take us beyond the scope of this book, one may be able to make a case for public–private coinvestment in worker training. In essence, governments subsidize worker training in firms, but firms will always bear part of the cost themselves, ensuring that government resources are generally directed to training programs that bring real benefits.

With a view to remedy this underinvestment in training, Tan, Gimpelson, and Savchenko (2008) suggest that the Russian government should consider putting in place employer-targeted training policies. ECA countries can learn from drawing on the experiences[28] of many other countries, both industrial and developing, that have used payroll-levy training funds, tax incentives for employer-sponsored training, and matching grants. They suggest that policies should be designed to increase competition in training provision from all providers, both public and private, including the employer. Further, they also cite the use of matching grants, which can help develop a training culture. The most successful schemes are demand driven, implemented by the private sector, and intended to sustain the markets for training services. With a view to generate training capacity in enterprises and increase the propensity for workers to undertake training, grants in ECA should aim at strengthening and diversifying the supply of training and stimulating demand.

The cost is only one barrier to effective worker training. Indigenous firms behind the technology frontier are often not knowledgeable about the kinds of training programs that could effectively equip their workers to manage new technology in an efficient manner. The Serbia case stud-

---

28. Such as those cited with regard to training levies in Middleton, Ziderman, and Adams 1993; Gill, Fluitman, and Dar 2000.

ies describe intensive training efforts conducted by foreign owners to bring the acquired firms to the technical frontier. In some cases, this included bringing assembly line workers and a shop foreman into established plants in other countries, so that front-line workers could receive direct advice and instruction from their peers in the parent company. Training manuals and training procedures used in contexts like this can often be considered strategic assets of a foreign firm; there may be a natural reluctance to share such knowledge with unaffiliated indigenous firms.

But there will be other circumstances in which such knowledge sharing may be in the mutual interest of local and foreign firms. As documented in East Asia, foreign buyers are often willing to share detailed technical knowledge with local suppliers, enhancing the worker training process. Foreign manufacturing firms located in ECA have strong incentives to ensure that direct and indirect suppliers meet quality and efficiency standards, and there will be incentives for knowledge sharing in those contexts as well. ECA firms and governments should make the most of the opportunities that this confluence of interests creates. And, of course, this line of thinking again underscores how important it is for the region's countries to continue to embrace trade and FDI openness.

How well do ECA firms absorb knowledge?

Having extracted significant lessons for the ECA region from the existing literature, we go on to use original data analysis to quantitatively assess their relevance for the ECA. The Business Environment and Enterprise Performance Survey (BEEPS) datasets are cross-sectional surveys. Its datasets contain a number of variables describing outcomes closely related to technology absorption, each of which is obtained from detailed surveys of managers of ECA-region firms. In these surveys, firm managers are asked specifically whether their firm recently introduced a new (to the firm) product, upgraded an existing product, acquired a new production technology, signed a new product licensing agreement, or acquired a certification to an international standard. Each of these potentially represents a dimension of the kind of technology absorption process that we believe is fostered through exposure to international best practices. We use these variables to attempt a systematic assessment of the degree to which ECA-region firms really are absorbing technology through their connections to the global economy specified in the academic literature.

We are able to measure this connection because the BEEPS data contain information on the extent of international "connectedness" of the individual firms. These data include information on exports, the level of foreign ownership, whether the firm is a supplier to a multinational, how much it sells to multinationals, and whether the firm engages in joint venture partnerships with multinationals. It is thus possible, at least in principle, to measure the impact of international connectedness along various dimensions on the likelihood of technology absorption, measured in different ways. Based on prior World Bank research, we know that the level of human capital within the firm and the institutional features of the home market that reward or penalize investment will also impact technology upgrading and technology acquisition. We will include these variables in some specifications to ensure that our key findings are robust.

What do the regressions tell us about the links between trade, FDI, and human capital and technology absorption? Our main findings are shown in table 2.2.

First, we look at technology absorption as measured by the introduction of new products. We find that *being an exporter, compared with being a nonexporter, is associated with a higher likelihood of the introduction of new products or processes* of about 6.5 percent. Moreover, an increase in the export sales ratio implies an increase in the likelihood of the firm introducing a new product or process. Viewed together, these findings suggest that there is a strong association between the engagement of a firm in export activity and measures of technology absorption, but that the exact level of export activity is a weaker predictor of technology absorption. In addition, we find that a joint venture with a multinational firm is associated with a higher likelihood of introducing a new product or process of more than 100 percent.

Second, *we see a similar pattern when we adopt an alternative measure of technology absorption: the upgrading of an existing product or process.* All but one of our measures of international connectedness—the percentage of sales to multinational corporations (which is negatively associated)—are positively associated with technology upgrading. There is a 5 percent increase in the likelihood of product or process upgrading, as one moves from nonexporter to exporter status, and a nearly 20 percent increase in the likelihood of product or process upgrading as one moves from the absence to the presence of a multinational joint venture.

Does this suggest a more limited technology transfer to wholly owned or majority-owned subsidiaries? Is majority ownership by foreign firms a handicap in the technology absorption process? That interpretation would be inconsistent with one of the most well-established empirical regularities in the micro TFP spillovers literature: the finding that wholly or partially owned foreign firms tend to be more productive than other firms in the industry. In fact, when we construct simple measures of labor productivity using the BEEPS data, we find similar evidence of a positive relationship between being majority owned by foreign firms and having higher productivity.

We interpret these results in a different way: affiliates of multinationals are likely to acquire a high *level* of technology soon after acquisition; hence, they are less likely to have to upgrade. This is consistent with the studies that find that foreign affiliates are more productive—that they have a higher *level* of technology and productivity.[29]

This particular aspect of the general pattern of our results should be quite reassuring for those who might be concerned that the benefits of trade and FDI openness are largely restricted to enterprises controlled by foreign investors. We seem to be finding just the opposite—majority foreign ownership does *not* appear to be strongly associated with our measures of technology absorption. Rather, it appears that firms controlled by *local* owners, engaged in exports, or participants in FDI-mediated supplier networks are the primary beneficiaries.

Third, we ask whether the firm in question has acquired a new production technology. Here, too, *all the measures of international connectedness, except majority foreign ownership, are found to be positively correlated with the acquisition of new technology.*

Fourth, we construct a dependent variable for absorption, "technology upgrading," from a composite of dummy variables measuring firms' introduction of new products or processes, upgrading of existing products or processes, achievement of new certification to international standards, and acquisition of technology licensing agreements from other firms. Once again, all measures of international connectedness, except majority foreign ownership, are positively and significantly associated with technology upgrading.

There is also a robustly positive relationship that exists between measures of human capital at the firm level and technology absorption. In the enterprise surveys we use, human capital is measured by the existence of in-house training, a ratio of the number of professionals to total employees, and a ratio of the number of university graduates to total employees. The results for training are particularly strong.

---

29. An alternative interpretation of this result is that majority foreign-owned firms may apply a different set of criteria in determining whether a particular technology is "new," and in determining whether a particular process or product is being upgraded. It is quite likely that the managers of foreign firms will either be foreigners themselves, or they will be indigenous managers with a very deep understanding of technological best practices as it exists in the global economy beyond the ECA region. Such individuals are less likely to flag a process or technology as "new" when it only brings the ECA enterprise up to standard practice elsewhere, whereas an indigenous enterprise manager might be much more likely to view adoption of the identical process or technology as "new." To the extent that our interpretation is correct, it suggests that empirical results regarding the impact of majority foreign ownership have to be viewed with care and a certain degree of skepticism.

TABLE 2.2
Openness is better: Link between international interconnectedness and technology absorption

Variable	Introduction of new products	Upgrading of existing products and processes	Acquisition of a new production technology	"Technology upgrading" variable
Exporting firm	+***	+***	+***	+***
Majority foreign ownership	–	–	–***	–***
Joint venture with a multinational firm	+***	+***	+***	+***
Size	–	–	+	+***
Age	–***	+	–***	–***
State ownership	–*	–***	–*	–***
R&D expenditure	+***	+***	+***	+***
Web use	+***	+***	+***	+***
ISO certification	+***	+***	+***	
Training	+***	+***	+***	+***
Skilled workforce	+***	+***	+	+***
Infrastructure index	+*	–	–**	–
Governance index	+**	+***	+	+***
Use of loan	+***	+***	+***	+***

* denotes significance at 10%; ** denotes significance at 5%; *** denotes significance at 1%.
Note: Number of observations: 7,964. All ECA countries except Turkey included in the sample. ISO = International Organization for Standardization.
Source: Authors' calculations; BEEPS 2002, 2005.

Further, the regression analysis controls for the impact of information and communication technology (ICT) and international standards on technology absorption. ICT is considered the preeminent "general purpose technology" of the past 20 years, as it has driven economywide growth over a range of sectors by prompting them to innovate and upgrade further, with technological progress in these sectors in turn creating incentives for further advances in the ICT sector, thus setting up a positive, self-sustained virtuous cycle.[30] Similarly, there is strong evidence

30. See Bresnahan and Trajtenberg 1995, Helpman and Trajtenberg 1996.

across a range of sectors that the adoption of industry standards is among the most important form of introducing product and process technology upgrades and increasing productivity for firms.[31]

So is there "learning-by-exporting"?

We believe that these results are useful, in part, because of the light that they cast on the debate over the existence of a "learning-by-exporting" effect. A long, well-cited series of case studies[32] has documented the importance of the process by which firms in East Asia learned to improve their manufactured products and manufacturing processes through their efforts to export to more advanced foreign markets. However, most attempts to identify positive TFP growth effects from learning-by-exporting in firm-level or plant-level data have rejected the hypothesis that a transition of firms into exporting is associated with an increase in TFP growth. Researchers have effectively concluded that learning-by-exporting effects do not exist.[33] In an ECA context, Commander and Svejnar (2011) have suggested that, controlling for foreign ownership, there is no independent effect of exporting on firm productivity. However, a recent paper by De Loecker (2010) concludes that most of the past research has in fact been biased toward rejecting the hypothesis of whether there is learning-by-exporting because it did not take into consideration past export performance in a firm's productivity growth.

Our results suggest a possible different conclusion from past research—one that may apply far beyond the ECA context. In our panel data, *transition* to exporting is positively and significantly correlated with *increases* in measured technology upgrading. We find this to be true even when we control for foreign ownership, human capital, and environmental factors affecting export climate. This is consistent with the hypothesis that exposure to foreign markets fosters learning, and our results suggest that this learning effect is not limited to foreign-owned firms. It is, of course, not inconsistent with the view that firms, as they seek to transition to exporting, will invest in upgrading their technology to make themselves more competitive in foreign markets. In other words, technology upgrading could also increase exports, but, to the extent that the technology upgrading was motivated in the first instance by the desire to compete in a foreign market, it still reinforces the policy implications we stress.

This finding could be reconciled with the absence of TFP growth effects in a number of ways. First, it is likely that the implementation of the learning obtained through export experience does not come for free—

31. See Corbett, Montes-Sancho, and Kirsch 2005.
32. Pack and Westphal 1986.
33. Bernard and Jensen 1999; Clerides, Lach, and Tybout 1996.

upgrading products and processes requires expenditures on foreign technology licenses, consultants, worker training, and new capital goods. Second, active competition with other new entrants (from the same developing country) into the high-income economy export market could limit the ability of any one producer to appropriate the gains from learning-by-exporting. It would clearly be in the interests of the downstream customers of these suppliers to encourage such competition. In the absence of firm-specific output prices and cost measures, this increased competitive intensity could, as we argued earlier, squeeze measured TFP effects to something close to zero. Third, part of the gains from successful entry into foreign markets could accrue not to the firms, but to the (initially) small number of talented managers within the local managerial labor markets capable of managing world-class production operations. Regardless of which of these theories, if any, is correct, the essential policy implication is the same. This strengthens the case for further movement toward an open trade regime and to the elimination of policies that penalize or undermine exporters. ECA governments seeking to promote technology absorption should eliminate export disincentives and pursue a policy regime that provides appropriate support for export activity.

National R&D systems require reform

Given that ECA's R&D institutes will play a critical role in the absorption of knowledge and technology, is it important to ask whether they are up to the job? The answer is, not yet. As the World Bank report *Unleashing Prosperity: Productivity Growth in Eastern Europe and the Former Soviet Union* notes: "Currently, many of the S&T (science and technology) resources are isolated both bureaucratically (in the sense that they are deployed in the rigid hierarchical system devised in the 1920s to mobilize resources for rapid state-planned industrial development and national defense), functionally (in the sense that there are few links between the supply of S&T output by research institutes and the demand for S&T by Russian or foreign enterprises), and geographically (in the sense that many assets are located in formerly closed cities or isolated science/atomic cities)" (Alam and others 2008). A disproportionate share of total R&D resources remains locked up in specialized state research institutes, in which there are no strong incentives to transfer commercially useful technology developed within the institutes to private entities in the ECA economies, which could put the technology to productive use.

Our results support these criticisms. Recall table 2.1, listing the top Russia-based organizations that have obtained patent grants in the United States, the world's largest and most important single patent jurisdiction. This list is dominated by the local research operations of foreign firms.

Despite their absorption of a large share of Russia's total R&D expenditures, the state-led research institutes produce fewer patents than commercial companies. Surely, these research institutes and private commercial enterprises are not strictly comparable, as the type of research they do is very different. However, this does not detract from the main message (and even reinforces it), that resources going to these research institutes could be put to much more productive use by supporting R&D in the private sector. In Russia, the share of researchers in the population, and the aggregate outlays of R&D in GDP, are comparable to those of Germany and Korea, and are far ahead of those of Brazil, China, and India. But these high levels of inputs do not translate into high value added per capita, with Russia lagging behind the OECD, as well as other large middle-income countries in R&D outputs, along with a relatively low number of patents and scientific publications per capita (Schaffer and Kuznetsov 2008).

What is true in Russia appears to be true more generally throughout the Commonwealth of Independent States countries, but less so in Eastern Europe. The traditional institutions conducting research have been criticized for their isolation from international technological trends. We find statistical evidence of this relative isolation in the citation patterns of U.S. patents generated by indigenous inventors, when they are compared with comparable inventions of comparable vintage generated in other parts of the world. These indigenous patents generally make fewer citations to the existing state of the art, and they cite narrower inventions. The traditional institutions have been characterized by the poor quality of their inventions. We see evidence of the limitations of indigenous inventions in the relatively small number of citations that these patents receive from patents granted subsequently. Moreover, the number of indigenous patents is low relative to the level of R&D investment, and the number of patents is not sharply increasing, as they have been in countries such as China and India.

All of these findings reaffirm the need for continued efforts to reform ECA R&D systems, and to complete the transition from the socialist-era model to a system modeled on global best practices that is more internationally integrated and market driven. The following chapter will examine these questions and propose reform options that governments can implement in the short term.

Case study: The role of FDI in helping Serbia acquire technology

▶ **Serbian case studies of manufacturing illustrate how a firm's ability to tap into the world technology pool depends on its capacity to take tough decisions and make large investments to bring in and modify imported equipment and technologies, and restructure production lines and organizational structures.**

▶ **The case studies show that exposure to foreign markets fosters learning, both in foreign-owned and domestic firms. Specifically, acquisition by a foreign investor can strengthen the incentives and firms' capacity to carry out efficiency-enhancing factor reallocation, restructuring, and investment.**

Why did the Serbian government believe that attracting reputable international investors to its 2001 post-Milosevic privatization program was not only necessary for economic modernization but also feasible, despite Serbia's postwar difficulties? In the 1990s, armed conflicts and the dissolution of the former Yugoslavia led to international sanctions, which interrupted the production relations of Yugoslav companies, which were well established in Europe, and caused international isolation of these companies from input sources, as well as the loss of markets. These events left a dire economic legacy, and it was clear that companies required FDI to allow them to regain their position in European markets and to replace the technologies that had by then become obsolete. These companies included 86 firms that were selected for tender privatization owing to their contribution to exports, importance for employment generation, and so on. The privatization program was able to sell these enterprises—which were a collective form of property ownership that was controlled by the employees—and this was the core of its success.

The ambitious program centered on Serbia's 2001 Privatization Law (amended in 2003), which incorporates international best practices and lessons learned from a decade of experience in other transition economies. The law stipulates three methods of privatization: *tenders* of large enterprises, offering to a strategic investor at least 70 percent of the shares; *auctions* of medium-size enterprises; and *restructuring* and subsequent tenders and auctions of a select group of large, presently loss making, but potentially viable, enterprises, or parts thereof.[34]

In the first five years, 1,407 enterprises were privatized through competitive public tender and auction procedures, with privatization proceeds reaching nearly 1.7 billion euros, and social and investment program commitments of almost 1.4 billion euros. The fact that 74 percent of all offered companies were actually sold is impressive considering the numerous challenges resulting from the legacy of social ownership.[35] There were also about 800–1,000 companies that were privatized according to the 1997 Privatization Law, prior to February 2001, when the post–Milosevic government took power. The 1997 law gave away 60 percent of a company's shares to employees free of charge, and 30 percent were offered for sale to the insiders at a deep discount and in installments. Many years later, the following questions arise:

- What is the effect of foreign ownership on technology absorption—that is, is there a difference in the absorption process followed by firms acquired by a local *versus* a foreign investor, for example, a multinational enterprise (MNE)?

- How does ownership affect corporate governance, and how does the latter, in turn, affect absorption?

- Was management and organizational change a prerequisite for the implementation of new investments and technology?

- What were the effects of the investment climate on mergers and acquisitions (M&A) and FDI, and what are the corollary effects for Greenfield FDI?

- What determined the foreign investors' relocation of R&D to and from Serbia?

34. The privatization analysis in this case study is based on the authors' case study material, which was included in Goldberg and others 2007 and the results of Goldberg and Radulovic 2005.
35. Republic of Serbia Privatization Agency, http://www.priv.rs.

To answer these questions, we undertook a study of the firm-level absorption process, focusing on product mix, production technology, management, and skills, all of which illustrate the critical role that foreign strategic investors play in acquired companies that need to cope with multiple challenges of absorbing knowledge to increase their productivity and profitability. The knowledge comes in many shapes—it can be embedded in capital goods, derived from learning-by-exporting, brought in by consultants or other knowledge brokers, or codified in intellectual property that requires licensing. The case study also illustrates the high financial and nonpecuniary costs of absorption, and it compares how strategic foreign investors, local investors, and insider-owners confront and mitigate the costs of absorption.[36]

To be more specific, we looked at eight large companies operating in the metal processing, household chemical, pharmaceutical, and cement industries. The selection criteria were that: industries, as well as company characteristics (especially the size of the firm), were useful for the *comparability* of results; company characteristics and the type of acquisition (especially the type of buyer) provided some *controls to test counterfactuals*; companies were privatized early on to ensure the availability of archival information (due diligence, post-acquisition monitoring); and sufficient time for key restructuring and investment decisions to have been implemented.

Our emphasis on the outcomes of acquisitions of productive assets in Eastern Europe is still relevant today, because the legacy of mass privatization has frequently resulted in insider control that prevents openness to change and hampers absorption and innovation. And our results should complement the findings from the econometric analysis of the BEEPS and patent surveys, which provide evidence that multinationals contribute to *indigenous* technological improvement decisions and have played an increasing role in regional patenting activity. Although we tried to structure this case study to gain meaningful and robust insights about the "black box" of absorption through FDI by relying on a combination of qualitative interviews and financial analysis, we also drew on an external quantitative check—the 2005 EU impact study of the Serbian privatization.[37]

---

36. A "strategic foreign investor" is one that operates a business in the same industry as the acquired firm and is purchasing the business assets with the intention of operating them. In a transition context, "insider-owners" refer to managers who bought ownership stakes in former socially owned and state-owned companies.

37. The impact study is based on a survey of 187 companies, 122 of which were privatized under the 2001 law and 65 under the 1997 law. It focuses on the differences between the two groups. Three of our case studies were privatized under the 2001 law and only one—Albus—under the 1997 law.

Post-acquisition results

What did our case study show? First, acquisition is synonymous with a robust increase in operating income and a concurrent reduction in the workforce, both of which show up clearly within the initial two- to three-year time period, with FDI acquisitions showing more pronounced changes overall. The combined effect is a substantial jump in operating income per worker, ranging from 24 percent for the household chemical company bought by a local investor, to 308 percent for the aluminum

FIGURE 2.9
Revenue and employment trends pre- and post-acquisition[38]

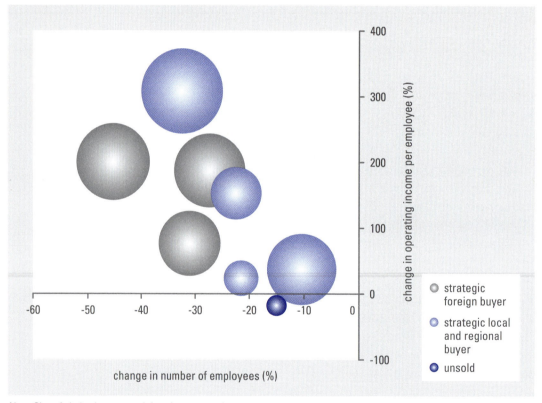

Note: Size of circle shows growth in salary per employee.
Source: Authors' calculations based on the annual Financial Statements of the companies 2001–06, Solvency Center, National Bank of Serbia.

38. In this and subsequent graphics/tables, the pre-acquisition averages are calculated from the financial statements from 2001 up to the year of privatization, which is different for each company; post-acquisition figures are inclusive of the privatization year and up to 2006. One of the main challenges of using these financial data was the change in standard accounting practices in these companies following privatization.

company bought by a foreign investor, with companies receiving FDI showing significantly larger gains. Salaries for those workers that remain following retrenchment and restructuring rise in line with efficiency improvements, and are between 99 percent and 170 percent higher for FDI acquisition companies. In stark contrast, the company that is yet to be privatized has seen a fall in operating income relative to the year when it should have been privatized, and salaries have stagnated.

Second, trends in firm-level productivity are generally consistent with the hypothesis that acquisition by a foreign investor brings about more efficiency-enhancing factor reallocation, restructuring, and investment, compared with local investors. The leading driving force behind higher labor productivity is downsizing of the workforce, as hiring and firing rigidities are removed by the introduction of new ownership arrangements. This is consistent with the view that privatization to strategic buyers permits labor shedding in enterprises that were subject to substantial labor hoarding. Our interviews with CEOs suggest that going forward, value added will rise much more as the companies turn from rehabilitating to purchasing new machinery and equipment, the lead time for installing new capacity is completed, and the effects of learning-by-doing show up in the financial data—a process that takes time. By comparison, the productivity of the nonprivatized household chemical company fell significantly, as overall value added declined in the same period.

Third, the financial results show changes in operating income, salary, and productivity of the same sign and about the same magnitude across the companies receiving FDI, though the way this story played out was different in each. The conditions on the ground had substantial impact on the decisions taken within each of the manufacturing plants—for example, the size and competition in the domestic market, the export potential for traditional product lines, and industry-specific best practices that are changing globally. The wide variation in restructuring and operating strategies is evident from the wide range of values for common financial ratios. One indicator that shows a more consistent trend is debt-to-equity, which increases far more for companies acquired by local investors, illustrating the point that they are more likely to obtain financing with domestic banks, rather than rely on equity injections and debt by acquiring MNEs in international capital markets.

Thoughts for policymakers

As for the policy implications, it is helpful to view them in terms of the initial questions that motivated our case study.

Foreign ownership and technology absorption

The financial results and interviews with the companies indicate that technology absorption depends on the initial incentives of the investors for making the acquisition. In companies bought by *domestic investors*, the motivation seems to be primarily horizontal or market-seeking—that is, winning a substantial share of the domestic market that can ensure sufficient profitability once the company is turned around. Domestic investors tended to repair and refurbish production assets and make targeted investments in new equipment to remove bottlenecks, rather than introducing totally different production and marketing methods. Their R&D strategy includes (passive) adaptation and (active) imitation of new foreign products launched in the market.

More radical changes in the product mix and manufacturing methods took place in companies bought by *foreign investors*, where the motivations included both horizontal and vertical (cost-minimizing) objectives. The frequent closures and switching of product lines indicate that MNEs took a regional or global perspective when rationalizing the product mix—one that accounts for the cost conditions of neighboring plants that could serve the same export markets. Economies of scale and access to markets are likely the main economic rationale for this process of regional or even global specialization. This determined the evolving product mix and, importantly, the weak R&D that has been financed after acquisition. In general, the research capabilities of the entering MNE are so advanced that only minimal domestic R&D for absorption is carried out, in contrast to the large-scale innovation of the MNEs at their R&D sites.

Are these results unique to the eight companies studied or do they reflect a broader reality? The EU impact study provides support for these findings. Taking a large sample of enterprises that were privatized, it finds that those companies privatized according to the 1997 law, and that came to be dominated by insider owners, on average have poorer financial results and performance. Their level of sales has not increased, even as average salaries have risen. Similarly, there are no signs of significant efficiency improvements or modernization efforts through transformational investments. The gradual changes that have been implemented imply that their aging production equipment and facilities will certainly worsen if no action is taken. However, those companies that changed ownership—that is, the employees' shares were bought by an investment fund or a strategic investor—show a marked advantage in performance compared with those with employee ownership.

We believe that by understanding the factors affecting what *investors* stand to gain from reducing the gap vis-à-vis the global technology frontier, it is possible to provide useful policy implications. The results show clearly that investment strategies of MNEs at the level of individual plants, companies, or countries are not formulated according to relative domestic costs alone but are contingent on past decisions and expansion plans for their nearby production facilities. Consequently, the potential for attracting strategic FDI through incentives will depend on regional and global variables that are not under the control of governments. One policy implication is that attracting FDI into certain industries has to be seen as a real race against other countries that can attract the same FDI partners. In this context, the saturation of local markets that are open to imports will lower the value of further investments, creating a tradeoff between liberalization and FDI promotion.

Because we know from the literature that concentrated ownership is important for corporate governance, and particularly for innovation and associated risk-taking, governments that want to stimulate innovation and change in state-owned enterprises that are slated for privatization should facilitate FDI through a properly regulated M&A process and, if still relevant, through good case-by-case tender privatization design. Such actions will increase the probability of attracting buyers with strong incentives to make transformational investments—though this could lead to less R&D in the near term as a byproduct. The dispersed ownership resulting from mass privatization has proven to be particularly problematic in postconflict countries such as Armenia, Bosnia and Herzegovina, former Yugoslav Republic of Macedonia, Moldova, and Tajikistan. In such circumstances, a strategic owner, local or foreign, is a *sine qua non* condition for good corporate governance, and consequently, for technology absorption.

Where there are insider-dominated, properly managed companies with the potential to attract FDI through M&A, the government could facilitate this process by consolidating more than 51 percent of the shares together with minority shareholders, which could be attractive enough to entice a strategic investor. In addition, in M&A the government could facilitate hiring of high-quality financial advisers for transactions, attracting a core (strategic) investor by accepting lower revenue for the sale of government shares, avoiding investment and employment commitments, and clearing past debts to the state so that firms can focus on forward-looking investments.

Ownership and governance

The introduction of modern corporate governance arrangements, where management is delegated control over most operating decisions by shareholders, who in turn have responsibility for monitoring and making key strategic decisions, was seen as very positive by all the managers interviewed.

A comparison of companies bought by local nonstrategic investors versus foreign strategic investors shows marked differences in terms of the sophistication of these arrangements and specifically the degree of separation between ownership and control. Simplifying for the sake of clarity, we could say that domestic-owned companies exert more direct supervision and control over operating decisions, and do so by a direct relationship between the owner(s) and top management. In the case of the foreign investors, rules and norms regulate reporting lines between the MNE and the subsidiary (for example, a matrix structure), and decision-making is tied to long-term planning methods that have to be agreed upon and adhered to.

The diversity of governance arrangements between investors and acquired companies suggests that the government may want to introduce rules to ensure minimum corporate governance after acquisition. Specifically, two measures that could be considered are the adoption and disclosure of corporate governance guidelines, and requirements about independent directors. Corporate governance is very relevant to our topic because we believe that it is a necessary condition to ensure incentives for risk-taking, which is a prerequisite for innovation and technology absorption.

Management

An essential part of successful FDI-driven absorption concerns the development of a competent managerial cadre with the appropriate incentives and tools. Managerial competences are essential for companies to plan and handle far-reaching changes in technology and workforce organization, and they need to be developed, exercised, and rewarded. This issue was identified in each of the cases, and could be highlighted as one of the triggers for a broad corporate transformation process that increases the value added.

Yet, there is no unique solution to the management dilemma—what Alfred Chandler (1977) appropriately called the "Visible Hand" in his path-breaking case study analysis of the managerial revolution in American business. In the companies we look at, some decide to replace all pre-acquisition management, others decide to keep most of the team; in some companies, the development of younger staff is paramount, but in others, the top managers are brought from outside, whether it be from the staff of the strategic investor or through head hunting from competitors. A common post-acquisition change is the introduction of a more powerful incentive structure for managers and workers.

This pattern is consistent with the results of the EU impact study. In the case of the privatization that took place under the umbrella of the 1997 law, only about half the companies have changed their managers since privatization. The percentage is much higher in the case of 2001 privatizations (close to two-thirds of those giving information on the status of their directors). In the 1997 cases, most new directors came from inside the company; in the 2001 cases, the majority came from outside.

Overall, the study finds evidence that managerial capacity is rapidly developed by buyers, and this process is often based on informal learning-by-doing of appointed managers as they interact with strategic investors. In the case of FDI especially, we conclude there is limited or no role for intervention by the government regarding post-acquisition managerial and organizational changes. However, because this case study is restricted to large manufacturing companies, public support could have a role to play for the healthy development of small and medium enterprises bought by investors with fewer resources and capabilities.

Organizational change

As for the workforce reorganization, in the companies that we examined, 20–50 percent of the workforce left the company after acquisition. This process has been accompanied by new hiring focused on sales and marketing, and the introduction of reward schemes to improve work incentives. For employees, the acquisitions had a mixed outcome: higher salaries and a better quality of employment were made possible by efficiency gains, though at the expense of a shrinking workforce. But in general the restructuring process and the wider effort to minimize costs were synonymous with a reduction in employment levels. Absorption through plant modernization and automatization depressed demand for labor for a given output, as the capital-to-labor ratio of new machinery and equipment tends to be higher.

At the same time, the arrival of fresh resources from the investor— and later on the achievement of a breakeven—allowed companies to pay wages and social contributions regularly, and to offer relatively generous severance packages that compensated workers affected by technical redundancy. Furthermore, the modernization of the companies directly improved the conditions of work by creating safer environments.

These results agree with the EU impact study, which found that privatization has had a negative short-term impact on employment, due to adjustments implemented by the companies. The substantial decrease of employment associated with privatization under the 2001 law was accompanied by a change in the qualifications of employees. Average salaries of companies privatized according to the 2001 law have jumped 130 percent the first year, to reach a total of 150 percent the second year. Companies privatized under the 1997 law did not make significant changes in employment.

In transition economies, the expected impact of technology and knowledge transfer elicited by FDI can be large because of the complementary technical skills embodied in an educated workforce. Without these, the introduction of new machinery and the implementation of quality certification would be unfeasible. The case studies show the importance of in-house postprivatization training programs for employees. We already mentioned that appropriability issues can be a barrier to the development of local managerial and technical capabilities. The same problem applies to the technical skills embodied in an educated workforce: high turnover deters locally owned companies from investing in in-house training (necessary to update existing skills) because they cannot run the risk of losing the newly trained employees.

Investment climate and FDI

In contrast to other Eastern European socialist countries, Yugoslav enterprises had a tradition of exporting to Western Europe. Serbia's comparative advantage in the 1980s was fruit processing from the northern region of Vojvodina and generic pharmaceuticals. As a consequence of the embargo imposed on the Milosevic regime in the 1990s, Serbian enterprises lost most of their export sales, and were only slowly rebuilding their network of customers when the 2001 privatization law came into effect. From an analytical perspective, this situation creates a "natural experiment," as we can observe the export profile prior to the embargo period and after acquisition. Every company starts from a position of forced autarchy, and once the external constraint ends, the management needs to decide how to serve foreign markets.

Our case studies suggest that reestablishing a presence in foreign markets without an alliance, joint venture, or FDI is a difficult undertaking. In general, the companies sold to domestic investors that we have examined (Albus, Nissal, and Zorka) have not been able to increase exports in such a significant way, while their comparators (Merima, Seval, and Zdravlje) are doing much better.

Relocation and R&D

In Serbian companies acquired by foreign investors, the comparative advantage for R&D lies in the adaptation of products and machinery to local conditions. For example, advanced formulas or product designs are transferred from the MNE and adapted locally, so that they can be manufactured efficiently in the acquired plant. There is also need for introducing minor marketing-led innovations, screening of competitor products, quality control, and establishing standards, among other activities. The underlying reasons for not maintaining large R&D facilities locally, as we pointed out, are that economies of scale and scope push toward a consolidation of innovation and production activities in large specialized facilities usually located in the same country as the headquarters of the MNE.

Policymakers need to be aware of the advantages and drawbacks of having manufacturing firms in which the owners have minimal incentives for innovation-seeking R&D, and which instead spend resources primarily on the task of transferring technology. On the positive side, this tends to accelerate the movement of the industrial base toward the global technology frontier, which is critical for increasing productivity in the short run, and increases the incentives for domestic rivals to upgrade. And technology absorption could be a necessary first step toward a more ambitious innovation agenda for the industrial sector.[39]

In Serbia, the large technology gap between foreign entrants and domestic consulting companies or domestic research organizations creates problems for collaboration, whether this regards consulting services or research consortia with local researchers. This weakens the effectiveness of supply-side government policies to promote technological progress, regardless of how well structured the policies are. For example, establishing a fund to encourage collaboration between research organizations and industry is unlikely to draw much interest from companies that have already received FDI and have advanced several steps on the technology ladder, because the subsidy cannot compensate for the transaction costs and delays involved.

To meet this challenge, the government can consider two courses. First, to concentrate its attention on creating incentives that support the absorption process by local industry. Second, to make the deep-seated changes and substantial investments required to restructure the old

---

39. European Commission 2007.

research and development institutes (RDIs), so that the public R&D infrastructure can play an active role in the industrial transition from absorption to innovation. As knowledge, commercial innovation, and R&D become a priority in ECA's advanced reformers, the industrial RDIs, inherited from the centrally planned system, have not been restructured in many ECA countries. Scarce resources spent on subsidizing industrial RDIs could have been used more efficiently to encourage innovation. In addition, the restructuring of industrial RDIs would stimulate the transition of applied R&D and laboratory workers to private enterprises. Restructuring would resolve some of the current intellectual property conflicts of interest created by the systemic moonlighting of RDI workers in private enterprises.

CHAPTER 3

Connecting research to firms— options for reforming the public RDIs

▶ In many OECD countries, RDIs occupy an important role in the national innovation system—clients often include small firms that lack the capability and market intelligence to identify their own technological, organizational, and managerial needs. However, RDIs in ECA have yet to complete the transition to the region's new economic realities.

▶ Because ECA firms fund little R&D, government R&D plays a much larger role: around 60 percent of all R&D in OECD countries is funded by industry and 30 percent by government, while in ECA countries, the proportions in financing are reversed. This explains why improving the performance of RDIs in ECA is a vital agenda for innovation and competitiveness policy.

▶ To connect research to firms and tilt the public–private R&D balance, irrelevant RDIs need to be restructured, closed, or privatized depending on the specific context.

▶ **Our recommendations suggest that:**

- Strategies that involve corporatization but retain government-ownership, insider restructuring, or insider privatization are politically feasible but least effective in governance reform.

- Strategies that entail outsider privatization and liquidation or closure are the most difficult politically but sometimes urgently needed to reduce wasteful public spending.

- Strategies in which RDIs remain government-owned but contractor-operated or are transformed into a nonprofit foundation have proved successful in the EU and the United States, but they may not be feasible in ECA because of corruption and regulatory capture concerns.

Research and development (R&D) institutes in Europe and Central Asia (ECA) are a legacy of central planning that is widely recognized as part of the unfinished restructuring agenda in ECA. During the socialist period, research and development institutes (RDIs) were part of a production and innovation system in which most often R&D was not performed in-house within enterprises and therefore was not directly driven by production needs or market demand (Radosevic 1998). The lack of a market for technologies or feedback mechanisms between end users and sources of innovation limited the diffusion of technology and innovation. The separation between the supply and demand for innovation was particularly visible in the Soviet Union where industry–research links were mediated by the responsible ministry. Funding for all these RDIs was provided directly or indirectly by the state. In Central Eastern Europe, particularly in the former Yugoslavia, the links between industry and research institutes were closer (Meske 2004). Although Turkey does not have the same history of central planning as other ECA countries, by its very nature, Turkey's RDI system has not been successful in establishing strong connections between research and commercialization activities.

Prior to the transition in the 1990s, ECA countries typically had three main actors comprising their science and research systems: the Academies of Science, the branch sector, and universities. Of these, the Academies of Science were dominant in their influence over the national science and technology system, primarily conducting basic or fundamental science[40] research and helping to establish national science policy. The Academies of Science had their own research institutes, and these typically focused on basic research. Turkey's RDIs were organized under the Scientific and Technological Research Council of Turkey in 1963 while others continued to operate under specific ministries.

The branch sector was the primary institutional home for applied R&D. The bulk of applied research in ECA countries pretransition was concentrated in the industrial RDIs, which fell under the purview of the industrial or branch ministries. The state owned these RDIs, and various industrial ministries supervised the activities of industrial RDIs located in their respective branches. The military or defense sector accounted for much of the applied research activity in transition countries, particularly in the former Soviet Union. During the Cold War, most of these countries developed extensive, state-funded, top-down, military–industrial complexes. Universities in ECA countries were mostly focused on teaching and conducted very little research. Poland is among a few exceptions where the share of R&D performed at universities, even in socialist times, was relatively high.

The overarching feature of ECA RDIs was the centralization of decision-making and administration, which led to little direct cooperation and communication among the various segments of national innovation system. R&D was not organized as an in-house activity or R&D in industry, but as R&D for industry (Radosevic 1998).

The restructuring of *industrial* RDIs in the transition has proved problematic as they could not be clearly classified as public or private organizations. Their knowledge profile overlapped public–private boundaries, which led to very different privatization approaches across ECA. Many ECA governments believed that by privatizing these institutes, enterprises would lose contact with organizations that were actually substituting for their missing R&D capabilities.

Incomplete restructuring of RDIs

The collapse of state funding and budgetary difficulties led to greatly reduced R&D funding in the 1990s. In addition to the lack of funds, the decentralization of decision-making and abolishment of central planning led to the hastened dissolution of many research units (table 3.1). RDI units experienced a drop not only in their public funding but also in their orders from firms. This reduced demand stemmed from a dismantling of

40. The term "basic sciences" refers to the classical disciplines of mathematics, chemistry, physics, and biology, among others. Such research may have no direct or immediate practical application or commercial benefit.

TABLE 3.1

A massive overhaul of RDIs in the 1990s

Country	Key features
Belarus	Explicit policy of preservation of R&D organizations irrespective of institutional sector.
Bulgaria	Industrial R&D institutes have gained autonomy and were left without state funding. Between 1996 and 2000, 150 branch R&D units were liquidated or privatized.
Croatia	Public R&D institutes have continued to be funded by public funds. There are no plans for the privatization of state research institutes. Enterprise R&D institutes have been either closed or downsized.
Czech Republic	Industrial R&D were immediately transformed into state limited companies and later privatized in two waves of voucher privatizations (109 institutes with workforce of about 30,000). Privatization of the enterprise-based R&D sector was completed in mid-1990s. Significantly downsized sector and transformed industrial R&D organizations has been reoriented toward market demand offering often non-R&D services.
Estonia	All of the 23 Soviet-era R&D institutes, except in agriculture and two in energy, have been closed. Research institutes of Academy (17) have been integrated into four universities. Seven state research institutes have retained that status.
Hungary	In 1992/93, industrial R&D institutes were transferred to the state privatization agency, but the state retained controlling share. Institutes have been left to their own devices and majority of them have closed. Only three have been privatized. Of 25–30 industrial institutes in the 1980s, only 3–4 remain today and are involved in non-R&D activities. FDI has established several tens of R&D labs in Hungary and several enterprises R&D units have been preserved and expanded.
Kazakhstan	Industrial R&D institutes have been converted into government-owned R&D organizations that operate based on ministry funding and market contracts. There has not been a policy of active restructuring of these organizations.
Latvia	Most Soviet-era R&D organizations have been either closed or left without core funding. Industrial science has been ignored in restructuring and funding. State research institutes not integrated into the university system will remain or become national research centres (centres of excellence).
Lithuania	Of 29 state research institutes, 13 have been merged into higher education institutions while 6 have become part of higher education institutions. Others have been retained as state research institutes but without core funding.
Poland	RDIs represent a government-run enclave in the economy (115 supervised by Ministry of Economy). Organizational and ownership transformation has not been undertaken. Privatization is voluntary. So far, none of industrial RDIs has been privatized. However, partial privatization of R&D units is common (50 percent of R&D units). Continuous state funding of RDIs enables them to carry out research without much regard for users needs. They have transformed themselves in hybrid organizations conducting R&D and commercial production or service activities.
Romania	Industrial RDIs were commercialized and gradually transformed themselves into non-R&D organizations or hybrid organizations, which combine a variety of activities. Some have been privatized.

Russian Federation	Partial differentiation among R&D organizations by establishing state research centres. Policies of preservation of R&D organizations gradually adjust by shifting to non-R&D activities. Policy influences their restructuring only by changing funding criteria.
Serbia	There is still no specific government policy on restructuring and privatization in R&D sector. Gradual reorientation toward non-R&D activities. Enterprise RDIs have been either closed or downsized.
Slovak Republic	Similarly to the Czech Republic privatization of R&D organizations during the first wave of privatization in 1992.
Slovenia	No inherited industrial institutes. Public research institutes have not been privatized.
Turkey	RDIs have been subject to limited restructuring and a number of management and funding reforms were made in the early 2000s to increase links between RDIs and the economy.
Ukraine	Unreformed R&D system dominated by policy of preservation of the overall R&D system.

Source: Adapted from Meske (2004) and Radosevic (1998, 2005).

the large enterprises. In the former Soviet economies, the demand of the military–industrial complex enterprises for civilian R&D was never high, and in the wake of transition it practically disappeared (Sedaitis 2000).

With the collapse of state funding, payrolls correspondingly shrank. Before the transition, ECA countries benefited from a highly educated workforce: an abundance of scientists, engineers, and other R&D personnel. But during the transition thousands of scientists and engineers emigrated or moved overseas. In Latvia, at the start of the transition there were 50 RDIs that employed roughly 13,000 people—this meant that almost 1 in every 100 economically active people was employed in an RDI. By 1993, there were only 13 RDIs, employing 887 employees. In Bulgaria, the number of researchers decreased from 15,000 to 10,000 over 1996–2004, and many researchers continue to leave Bulgaria for opportunities abroad today.[41] Moreover, industrial RDIs needed to change their portfolio of activities to seek new sources of revenue such as contract research, technical consulting services, and even small-scale production. These non-R&D activities grew in the industrial RDI's share of total activity. But there has been limited integration of industrial RDIs into the manufacturing sector even as industrial RDIs became hybrids—part R&D unit, part commercial entity.

RDIs outside ECA

In many developed countries, RDIs occupy an important role in the national innovation system, providing support to industry, drawing from national and international industrial and scientific knowledge. These

---

41. UNESCO *Science & Technology Statistics*.

RDIs acquire, maintain, and supply technologies and technology-related services to firms that cannot access them in-house or from private providers. In some newly industrialized economies, they were initially used to address existing gaps in the national innovation systems. These include the Korea Institute of Science and Technology (KIST) in the Republic of Korea, the Industrial Technology Research Institute (ITRI) in Taiwan, China, and Hong Kong Productivity Council (HKPC) in Hong Kong SAR, China (Arnold and others 1998). In contrast to universities and basic research institutes, RDIs complement "hard" technological development activities with "soft" activities such as testing, troubleshooting, consultancy, training, seminars, and standards and certification.

Public RDIs in Western Europe are argued to play a "crucial role" in helping to overcome "market failures" and capability failures. Research associations originally tackled common problems within one or more branches of industry and then became institutionalized in the form of institutes (for example, Mekanförbundet, a branch organization of the Swedish engineering industry, which eventually established the Swedish Institute of Production Engineering Research). These public RDIs provide small and medium enterprises (SMEs) with independent expert advice that can help bridge the information gap that is at the heart of this market failure (Arnold, Javorcik, and Mattoo 2007).

Whereas RDIs in ECA were oriented toward the technological needs of large state-owned enterprises, RDIs' clients often include small firms that lack the capability and market intelligence to identify their own technological, organizational, and managerial needs (capability failure). However, supporting this market segment requires specific skills in marketing and business that many universities and research institutes are unlikely to have. Many RDIs, such as Centros Technológicos in Spain and HKPC, are explicitly organized to serve this market. Moreover, the SME market is typically very fragmented and rife with market failures, so even successful market-driven RDIs rely on government programs to support SME demand.

According to Arnold (2007), in most Organisation for Economic Cooperation and Development (OECD) countries, state-owned institutes have been transformed (or were already) into free-standing organizations outside the civil service. In particular, the separation between funding and research performance (or "customers" and "contractors" in the language of the influential Rothschild report which came out in 1971 and influenced reform policies in the 1980s[42]) is largely complete.

Although universities and research institutes increasingly need to form tight intellectual partnerships to meet the needs of industry, their

---

42. Rothschild 1971; Dainton 1971.

roles as service providers remain complementary. RDIs are easier and less risky to collaborate with because they are much more focused than universities and often use management processes and norms of confidentiality found in industry. RDI staff is often more experienced and familiar with practical production processes than in universities and can also deliver directly applicable knowledge to industry. Most important, providing services to industry is the core business of an RDI, whereas universities need to balance the tensions between teaching and research (Arnold 2007) and between "open science" norms that depend on early publication of research results and industry's needs for confidentiality and intellectual property right–protection (Dasgupta and David 1985, 1994; David 2004). However, universities are not constrained by the stability and consistency of institutes, which provide them with attributes that RDIs do not have. They are constantly regenerating their capabilities, with direct access to the next generation of scientists, and they are always under pressure to maintain a competitive edge to secure research grants. With this in mind, firms tend to rely on universities for human resources or for risky projects requiring strong problem-solving skills.

In general, RDIs in Western Europe and East Asia that have been credited for contributing to economic development—whether VTT (Technical Research Centre of Finland), Fraunhofer, ITRI, or KIST—have shared as a main characteristic that they do not attempt to substitute firms' innovative capacity but to complement it. Namely, their success is not based on their ability to invent a technology in-house and "sell" it to the market ("technology-push") but to provide solutions that respond to specific customer needs ("market-pull").

Given that innovation is not a linear process, the technology-push–market-pull dichotomy is in fact a simplification of the dominant R&D strategy approach. In reality, innovation in RDIs is likely to be a much more complex process involving technological trajectories, learning, and feedback processes from multiple sources. For strictly illustrative purposes, table 3.2 provides some of the characteristics likely to emanate from following pure technology-push or market-pull strategies.

The structure of government funding plays a key role in an RDI's market-orientation. When funding is provided as a subsidy with no strings attached, it can be used to pursue R&D objectives motivated by scientific curiosity—theoretically free of any commercial or political considerations—and can help build the long term R&D capabilities of an RDI. The RDI is likely to use this "core" or "strategic" funding to adopt a technology-push approach, where resource allocations are mostly motivated by internal scientific interest and capabilities. When this model is not accompanied by external performance evaluation, it can result in a lack

TABLE 3.2
A successful strategy typically reflects market needs

	Technology-push (Top down)	Market-pull (Bottom up)
Origin of the project	An idea from a scientist	A market need
Main barrier	Selling the idea	Identifying the market need
Technological sophistication	High to medium	Medium to low
Market	Unknown	Well known
Level of risk	High	Medium to low

Source: Nicolaon 2008.

of accountability of the RDI, poor performance, and lack of relevance with the economy's needs.

At the other extreme, if no income is available from the government, the RDI's survival will be based on offering services of value to the private sector. In this case, the RDI is forced to follow a market-pull approach, where its activities are solely determined by client contracts. A drawback is that to remain financially viable, the RDI will shift its focus away from R&D toward services for which its clients can fully and immediately appropriate payoffs. While still economically relevant, the RDI will lose its social role as a disseminator of public knowledge and in conducting research with high spillovers.

Status of reform efforts elsewhere

European Union. Reforms in European RDIs are diverse but share many common themes in governance reform: increased role of stakeholders, professionalization of management, changes in organization to become more outward-facing, increased autonomy to define and implement strategy, "contractualization" of relations with founders or customers through various kinds of performance contracts, often accompanied by performance indicator systems, and increased external quality control through the market (Arnold 2007).

United States National Laboratories. The most radical reforms in the U.S. RDI system took place in the 1940s. In the United States, the Department of Energy Laboratories are government-owned but are operated by private contractors selected from industry, academia, and university consortia. This government-owned, contractor-operated (GOCO) approach to laboratory management began in the 1940s to meet pressing wartime needs, and today provides flexibility in the assignment of resources and facilitates quick responses to a wide variety of program needs. This approach enables private sector and university-based R&D management experience to be brought to bear on government work. The GOCO system has offered significant advantages in attracting and retaining world-class scientists and achieving scientific excellence. In recent years, the GOCO system has been the subject of significant concerns regarding administrative and business management issues. At the same time, however, there has been a growing recognition that the GOCO approach utilized by the Department of Energy had resulted in generally superior technical performance than is found at government-owned, government-operated (GOGO) facilities.

Jaffe and Lerner (2001) examined the commercialization of publicly funded research in the U.S. national laboratories. Their results suggest that the organizational structure of the GOCO model used at Department of Energy laboratories may be far more credible than critics have suggested. Their empirical and case study analyses suggest that the policy reforms of the United States in the 1980s had a positive effect on technology commercialization with patenting activities sharply increasing and little evidence of degradation in patent quality. These effects appear to be stronger where the danger of bureaucratic interference was lower, such as when there was turnover of contractors. According to the authors, the results are consistent with numerous studies of privatized firms (D'Souza and Megginson 1999) that show transfers from public to private ownership to have a significant impact on performance.

China. RDI reform in China has been underway since 1999.[43] All RDIs were government-controlled, and the emphasis was on military technology. Research and production under central planning were disconnected, and there was little innovation in the RDIs. The objective of the reform was to enhance the development of the national innovation system, reinforce the role of enterprises in innovation, and facilitate the development of high-tech industries. The method undertaken was as follows: Research institutes were integrated into enterprises or corporate groups, or converted into high-tech enterprises or intermediary organizations.

By 2001, 300 centrally owned RDIs and more than 7,000 local RDIs were transformed into companies. The government made transformation into companies mandatory and maintained supportive policies for transformed RDIs: continued subsidies, tax holidays for five years, and pay-

---

43. Tang 2008.

ment of pensions. The majority of the equity is still held by the government, but notably some is owned by private shareholders or even listed on the stock exchange.

A survey commissioned by Ministry of Science and Technology in 2006 shows many positive results from these reforms: the staff size of enterprises converted from research institutes remains stable with a slight increase; the quality of new employees is better than that of leaving employees; employee income grows steadily; R&D investment has increased; the strength of the organizations has been enhanced; innovation capacity has been boosted; the pace of commercialization has been accelerated; financing and investment channels have been expanded; income from commercialization has increased; and their economic benefits have also improved greatly.

The remaining problems of the RDIs stem from a conflict in the funding system that leads to too little attention to long-term research. There is a dilemma for RDIs providing public goods: Should they become companies providing services or should they become manufacturing plants? And there is also the question of whether there should be preferential policies for those RDIs producing public goods and services.

India. The Council of Scientific and Industrial Research (CSIR) was set up in 1942, modeled after the U.K. Department of Scientific and Industrial Research. After India's independence in 1947, it focused on building up an extensive R&D infrastructure, from metrology to R&D for a wide range of industries—with a focus on supporting emerging industry, especially SMEs.

The reform process was initiated in 1986 and given additional impetus when India shifted from an inward-oriented to a more outward and market-driven development strategy as a result of the 1991 economic crisis. With the liberalization of trade and industrial policy, firms began facing more international competition. CSIR was criticized for being unwieldy and ineffective at transforming laboratory results to technologies for industrial production and for spending too much effort "reinventing the wheel" by focusing on known processes. The demands of the crisis led to self-examination and radical change in CSIR's role—from emphasizing technological self-reliance to viewing R&D as a business and generating world-class industrial R&D. More emphasis was placed on outputs and performance, along with work that was relevant for productive sectors and could earn income. Dedicated marketing and business development functions were established in each laboratory, which became a corporate subsidiary, and rewards were introduced for meeting targets. Laboratories were given autonomy in operations based on how well they delivered on committed deliverables. An external performance appraisal board was introduced to review the laboratories' performance every three years.

Although CSIR is still restructuring, the results to date have been quite impressive. They show the kind of impact that a change in the direction and incentive regime can have, even in a very large public research system. Between 1997 and 2002, CSIR cut its laboratories from 40 to 38 and staff from 24,000 to 20,000. Technical and scientific publications in internationally recognized journals jumped from 1,576 in 1995 to 2,900 in 2005, and their average impact factor increased from 1.5 to 2.2. Patent filings in India rose from 264 in 1997–98 to 418 in 2004–05. Patent filings abroad quintupled from 94 in 1997–98 to 500 in 2004–05, and CSIR accounted for 50–60 percent of U.S. patents granted to Indian inventors. In addition, CSIR increased earnings from outside income forms Rs 180 crore in 1995–96 to Rs 310 crore in 2005–06 (about $75 million). Today it has 4,700 active scientists and technologists supported by 8,500 scientific and technical personnel. Its government grant budget has roughly doubled since 1997, and is now Rs 1,500 crore ($365million), with earnings about 20 percent of its grant budget (Dutz 2007).

A snapshot of RDIs in ECA

Because firms perform so little R&D in ECA, government R&D has a larger role to play in the national innovation system than in the OECD. In OECD countries, 63 percent of all R&D is funded by industry and 30 percent by government. In ECA countries, the proportions in financing are reversed: 30 percent industry and 60 percent government. What is the status of RDIs in ECA countries today? What role do they play in providing R&D services to industry? And what challenges do they face? Evidence from the Community Innovation Survey suggests that cooperation with industry remains lower than in most EU15 countries.

To shed light on these questions, we undertook case studies of 21 RDIs in Croatia, Lithuania, Poland, the Russian Federation, Serbia, Turkey, and Ukraine. The information was collected through face-to-face interviews and informational material distributed by the RDIs or on their websites. For a sample of RDIs, data were also obtained through questionnaires covering a variety of topics, including missions and activities, scientific and market outputs, governance, management and funding models, and key assets. The authors' familiarity with the national innovation systems of these countries ensured that contextual information was not lost in the interpretation of the questionnaire and that national factors could be accounted for. Because the sample we used is not representative of RDIs in each country or of the ECA region as a whole, prior familiarity with the national and institutional context provided the authors with greater ability to interpret the data. In some cases, the RDIs in the sample are benchmarked against RDIs in a number of comparator countries in East Asia, Europe, and North America for which data are available through their annual reports, websites, or past RDI benchmarking studies.

Several of the RDIs included in the sample played a central role in their respective national innovation systems prior to the transition period (all but one RDI in the sample were created before the transition) and continue to be among the publicly funded institutions with a better track record. We chose to look at institutions that were originally better funded and closer to the technological frontier as this helps pin down inherent limitations of the gradualist strategy to strengthen the institutions. At the same time, to examine policy options for a broad range of institutional features, special care was taken to include RDIs spanning various typologies. In terms of size, 18 of the 21 RDIs in the sample are mid-size organizations, with between 100 and 500 staff members and most focus on highly specialized technical fields. The remaining three have between 1,000 and 2,000 staff. It must be noted that some of the benchmark RDIs from outside of ECA tend to be much larger, Foundation for Scientific and Industrial Research (SINTEF), VTT, the Netherlands Organization for Applied Scientific Research (TNO), and SRI International have between 1,900 and 4,000 staff members, while Arsenal has 180, and the Spanish Technology Centers—which are institutionally independent and are organized around a government agency, Fedit—have an average of 95 staff (table 3.3).

So what did we learn? Our findings show that two decades into the transition, the RDIs that are still operating as standalone entities have made limited progress in terms of the intensity and quality of their interactions with the overall national innovation system and specifically in the range of services they provide to industry, and it exposes areas that lag far behind such as knowledge management, licensing, incentive structures, and staffing.

Although the remaining RDIs are mostly owned or operated by government, two different types of RDIs appear to have emerged in regards to core activities and sources of funding. On one hand, there are RDIs that are predominantly funded by public sources and that are rather isolated from knowledge commercialization activities yet at the same time have not shown sufficient results in regards to publications and training and more generally fulfilling their mission to generate public knowledge with significant productive spillovers. On the other, some RDIs are largely financed through the goods and services they offer the private sector, but

TABLE 3.3
Foreign comparator RDIs vary in size and ownership

RDI Name	Country	Ownership	Number of staff
VTT Technical Research Center	Finland	Public	2,740
SINTEF	Norway	Public	1,901
SRI International	United States	Private	1,500
Fraunhofer Society	Germany	Public	239 per institutes on average
Arsenal Research	Austria	Private	178
Fedit Technology Centers	Spain	Public–private	95 per technology center on average
TNO	Netherlands	Public	4,003
PARC	United States	Private	230

Sources: 2008 World Bank survey of RDIs; RDI annual reports.

these goods and services seem to be at the lower end of the knowledge value chain, and there is a need to engage further in knowledge-intensive R&D activities that are competitive globally.

The foregoing deficiencies seriously reduce the effectiveness of RDIs as a mechanism for cutting-edge knowledge and innovations to be integrated into productive uses. The experiences of comparator RDIs from OECD countries suggest ways to improve this, such as introducing governance and funding models that achieve a better balance between strategic R&D and commercialization activities, reducing inherent tensions in the mission and organization of RDIs, and leveraging-in additional resources. Substantial policy challenges need to be faced to amplify the benefits of public investments in RDIs for the countries covered by this book to become part of the advanced club of knowledge-based economies, in which science, technology, and innovation lie at the forefront of productivity growth and competitiveness.

The RDIs in the sample differ from one another in their scope and focus of activities and span a wide range of industries. While some RDIs are engaged in fields that draw highly on scientific knowledge, such as biology, others specialized in technical areas that draw on a combination of industrial and scientific knowledge, such as shipbuilding, civil engineering works, and production of industrial equipment. Only two RDIs in the sample were engaged in several distinct technical areas (table 3.4).

TABLE 3.4

Specializations of the RDIs in the ECA sample

Technical area	Bio-1	Bio-2	Bio-3	Biochem	Chem	Elec-1	Elec-2	Energy-1	Energy-2	ICT	Mech-1	Mech-2	Mixed	Nuclear-1	Nuclear-2	Nuclear-3	Occup	Physics	Ship-1	Ship-2	Space
Biological and medical sciences	√	√	√	√									√								
Chemical engineering					√																
Chemistry													√								
Civil engineering											√								√		
Electrical engineering						√	√										√				√
Energy								√	√				√						√		
Environmental engineering											√								√		
Environmental science													√								
Food science													√								
ICT										√			√								
Mechanical machinery											√	√									
Nuclear engineering														√	√	√					
Occupational health and safety																	√				
Physics																		√			
Security and defense																			√		
Shipbuilding																			√	√	
Space																					√

Note: ECA RDIs have fictitious names for confidentiality purposes.
Source: Authors' analysis based on 2008 World Bank survey of RDIs.

For one RDI, part of this diversification was driven by market pressures to seek revenue in more profitable areas. For the other, different activities were put together to gain economies of scale in the administration of the RDI. It is worth noting that many of the RDIs in the sample do not fit neatly into the OECD notion of an RDI because they engage in production activities that are generally not covered by OECD RDIs.

What did the case studies show? Overall, the form and function of RDIs in the ECA are a legacy of the pretransition period. A few general characteristics stand out.

Limited resources. Funding constraints restrict upgrading of laboratories and the development of new R&D areas that respond to evolving industry needs. RDIs in the sample have far fewer financial resources per staff than their OECD counterparts. If the income is adjusted for purchasing power parity to account for the fact that salaries are a large share of expenses, the adjusted incomes of several RDIs are comparable to what one would find in OECD RDIs such as SINTEF and TNO. But a third of the RDIs have an income, which even when adjusted, accounts for a fraction of what one would see in OECD RDIs. These RDIs are highly under-resourced, which reduces their capacity to retain and attract skilled personnel, a point we discuss further below.

Little private sector involvement. In the sample, there are no fully private RDIs, either for-profit or not-for-profit. Of the 21 RDIs, 17 are state-owned institutions and 4 are joint-stock companies, owned by both the private and public sector. In the joint-stock companies, the state tends to own an overwhelming majority of the shares, mostly through a ministry or in some cases through a national academy of science.

This reflects a general trend found in the ECA region, where privatized RDIs, or their R&D activities, were not able to survive the pressures of an open, competitive market, largely because of industrial restructuring process that left little private sector demand for the RDI's services. By contrast, in OECD countries, private RDIs do exist, but they typically have some specific characteristics. They either conduct work of an applied nature and are not even designated as "RDIs," but rather engineering companies or design consultancies in their home countries, they are part of larger corporate groups, or they operate on a not-for-profit basis.

Not very market-oriented. With the exception of selected joint-stock RDIs, most RDIs do not appear to be very oriented to the market. The RDIs in the sample are engaged in a range of activities along the innovation chain. While some are mostly engaged in strategic R&D with no immediate commercial applications (Bio-3), at the other end of the spectrum, some conduct little original R&D but are oriented toward client-work with immediate practical applications (Energy-1 and Elec-2). In the sample, the former category tends to be overwhelmingly fully state-owned, while the latter has a greater representation of joint-stock RDIs. Nonetheless, several fully state-owned RDIs conduct very little strategic research (ICT, Bio-2, Ship-2). This appears to be a trend in many ECA countries, where a number of state-owned RDIs have phased out their strategic research during the transition period. It contrasts with OECD countries, where strategic research is the core learning and knowledge-generation mechanism that is associated with the types of spillovers of publicly owned RDIs.

A fuzzy teaching-research link. A number of RDIs are involved in teaching activities, but their role in bringing teaching and research closer together is often unclear. In ECA, under the central planning system, there was little overlap between teaching and research, with teaching the exclusive domain of universities and research the exclusive domain of RDIs. One exception was doctoral-level programs offered by RDIs. In some countries, RDIs have taken steps to formalize their teaching roles. The Bulgarian Academy of Sciences has established a nationally accredited graduate school that allows graduate students to be involved in its extensive research activities. It currently enrolls 554 graduate students and graduates 20 percent of Bulgaria's PhDs. In Russia, as of January 1, 2008, the integration of research and education is further simplified due to additions to the Federal Law "On Science" according to which RDIs and universities may jointly use material resources and workforce for research and teaching; facilities may be shared free of charge; and RDIs and universities may jointly establish integrated entities—such as educational centers and laboratories.

Limited transfer of scientific and technological outputs. ECA's RDIs appear to be falling behind in generating new forms of knowledge as well as transferring existing knowledge to industry. Indicators of knowledge generation such as publications and patenting are as equally mediocre as indicators of knowledge commercialization such as licensing activity and industry contracts.

When we look at publications in leading international journals (compiled using the Science Citation Index[44]), the performance of ECA RDIs is very weak, whether this is measured by the frequency of publications or the impact these have on the wider scientific community. The number of publications per RDI worker in a given year, which is a common proxy of the productivity of RDIs in generating and disseminating theoretical knowledge, is either null or very low relative to comparator RDIs located in the OECD (figure 3.1). Publishing in international peer-reviewed publications is either not valued in most ECA RDIs of the sample or the quality of the research is not high enough. Moreover, even those RDIs in the sample with a high level of strategic research activities and with few or no customers publish very little.

While a number of the RDIs in the sample are involved in patenting, virtually none of their patents led to licensing revenue. But the sample reveals a mixed bag. While some are extremely successful at patenting, others perform poorly. Some of the RDIs in the sample are engaged in providing services, which, though they rely on R&D, result in incremen-

---

44. The dataset in the Science Citation Index tracks the citations of all major publications and leading science journals. Arguably, this index is skewed toward the more prestigious and well-known publications, thus omitting the citations in less regarded local sources. But because this index represents the forefront of international scientific efforts and is comparable to a greater degree between institutions, it provides a more realistic picture of the relative distance to the knowledge frontier.

FIGURE 3.1
Number of annual publications per RDI staff

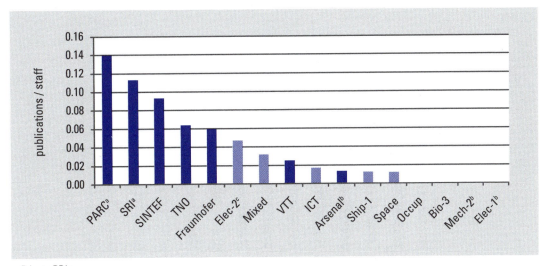

a. Private RDIs.
b. RDIs for which the state owns less than a 90 percent share and the private sector owns the remaining share.
c. RDIs for which the state owns at least a 90 percent of share and the private sector owns the remaining share.
Note: Fraunhofer, Mixed and Space based on 2008 data; all others based on 2003–07 averages.
Source: Authors' calculations using Science Citation Index.

tal improvements and adaptation of existing technologies, and patenting does not fit in their business model. This is also the case of some industrially oriented RDIs in OECD countries.

Licensing does not play an important role among the RDIs in the sample, and most have not licensed any technology in the past five years. Only one RDI had generated any substantial licensing revenue during that period. This is surprising when the RDIs are involved in technical fields in which licensing is a strategic mechanism for transferring technology, such as the biological and medical sciences. Of the three RDIs in the sample involved in those areas (Bio-1, Bio-2, and Bio-3), none had developed any non-negligible licensing activities in the past five years or generated any licensing revenue in 2007. All three RDIs were relatively successful at publishing their results, indicating that the absence of licensing is not the result of low research quality, but rather of deficient knowledge management, poor links with the private sector, or simply poor demand for their technological outputs.

Developing quality-related services such as testing and certification could help RDIs diversify funding sources and acquire experience working with industry. An examination of sources of revenues by activity indicates that several RDIs in the sample derive a substantial share (20–30 percent) of their income from testing, certification, and measurement services (figure 3.2). These are the types of practical commercial services

FIGURE 3.2
R&D and technical services to industry mostly marginal compared with public funds

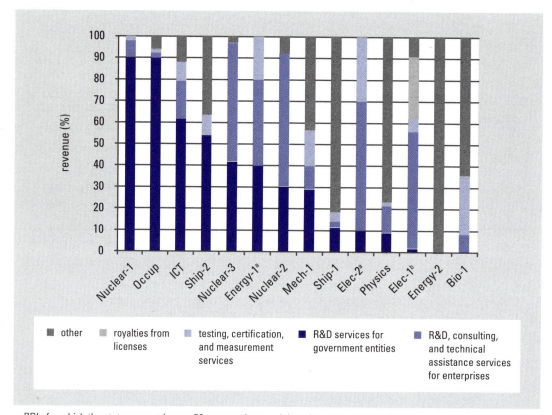

a. RDIs for which the state owns at least a 90 percent share and the private sector owns the remaining share.
b. RDIs for which the state owns less than a 90 percent share and the private sector owns the remaining share.
Source: Authors' calculations based on 2008 World Bank survey of RDIs.

that many RDIs in OECD countries focused on before they developed their science-intensive activities. Although not very scientifically rewarding for researchers—and not always recognized as central to the national innovation or research strategy—it is precisely these routine services that can be integrated in the RDI's overall business strategy because technical capacities and instrumentation are common to both activities. They can first and foremost act as a continuous stream of relatively secure income. They also can create business relationships with a potential industrial client base for more in-depth R&D services.

The joint-stock RDIs in the sample are more successful at forming business ties with the private sector, but SMEs are generally absent from the picture. The sample shows that the joint-stock RDIs in ECA are able to generate more income from industry than their public counterparts. The best performing RDI is Elec-1. However, it is worth noting that a large share of Elec-1's services are sold to other firms of its holding group, with which it benefits from long business and ownership relationships,

FIGURE 3.3
Some ECA RDIs generate as much industry revenue as international benchmarks, but this is not the norm

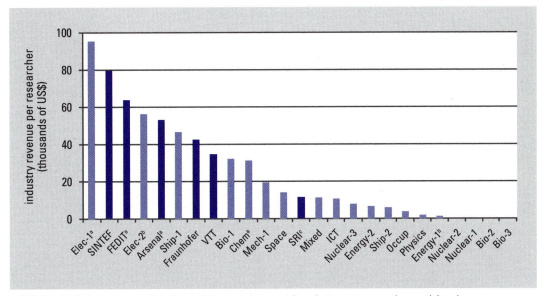

a. RDIs for which the state owns less than a 90 percent share and the private sector owns the remaining share.
b. RDIs for which the state owns at least a 90 percent of share and the private sector owns the remaining share.
c. Private RDIs.
Source: Author calculations based on 2008 World Bank survey of RDIs.

much like the relationship between Xerox and its Palo Alto Research Center laboratory based in Silicon Valley.

Some of the ECA RDIs in the sample are as successful as their Western European counterparts SINTEF, Arsenal, and VTT in generating industry revenue, but the top performing RDIs appear to be performing services that are better described as routine engineering services than R&D (figure 3.3). At the other extreme, two RDIs in the sample engaged in biological sciences have no revenue from industry. This could be a symptom of low productivity of the RDI, lack of marketing efforts, or lack of relevance of their research to industrial applications.

Moreover, the RDIs in the sample obtain a negligible share of their income from international businesses, even for those RDIs operating in small markets. The amount of business RDIs are able to win from international firms is an important measure of competitiveness in ECA countries. International business ties implies that the RDIs are not protected by national procurement preferences or by national monopoly positions but are competing on an equal footing with RDIs in other countries on the basis of price and quality.

In addition, few RDIs are being tapped for their technological capabilities by SMEs. Less than half the RDIs in the sample have SMEs as a

FIGURE 3.4
SMEs could make greater use of RDIs

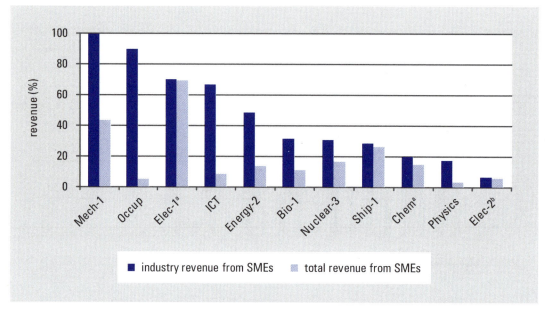

a. RDIs for which the state owns less than a 90 percent share and the private sector owns the remaining share.
b. RDIs for which the state owns at least a 90 percent of share and the private sector owns the remaining share.
Source: Author calculations based on 2008 World Bank survey of RDIs.

principal source of industrial revenue (figure 3.4). In the other cases, SMEs account for less than 20 percent of revenues. This is as expected since, as a general rule, SMEs—and particularly in ECA—invest less in R&D than larger firms. Nonetheless, while RDIs in OECD countries increasingly offer specialized services to SMEs, this is still a relatively uncommon phenomenon in ECA.

Government funding and governance

Are government funding and governance policies on the right path to revitalize the RDIs? Unfortunately not. Our assessment shows that greater competition among RDIs and a stepped-up private sector role are needed to help turn the situation around.

Funding policies are not conducive to effective market-oriented R&D.

The RDIs in the sample are associated with a variety of funding models (figure 3.5). Some institutions draw the quasi-totality of their funding from government subsidies (such as Nuclear-2, Bio-3, Occup, and

FIGURE 3.5
A bias toward a few types of funding sources

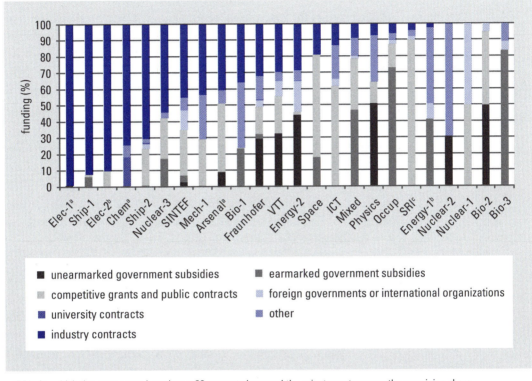

a. RDIs for which the state owns less than a 90 percent share and the private sector owns the remaining share.
b. RDIs for which the state owns at least a 90 percent of share and the private sector owns the remaining share.
c. Private RDIs.
Source: Author calculations based on 2008 World Bank survey of RDIs.

Energy). In one case, a joint-stock company receives almost 80 percent of its funding from government subsidies (Energy-1). Two of the RDIs in the sample receive the vast majority of their funding from core funding, with very little from competitive funding. This is a pattern repeated in many ECA countries. In Bulgaria, competitive funding only accounted for 20 percent of government funding for research in 2007. However, some European Union (EU) accession countries have moved in the direction of OECD countries. By 2006, only 13 percent of Slovenia's national research budget was allocated through institutional transfers, the rest being competitive funding allocated mostly through the Slovenian Research Agency.

But some RDIs in the sample derive almost all of their income from industry. These include most of the private or joint-stock companies of the sample. Five RDIs in the sample derive more than 70 percent of their income from industry. It is surprising that two of these five RDIs are still fully owned by the government. They are either providing services that have a limited public goods element, or are active in providing high spill-over services on the basis of strategic assets acquired through prior government funding. In the former, the logic of public ownership is not justified, and in the latter, the absence of government funding cannot be justified if the RDI is to maintain its existing function without seeing all of its equipment and knowledge become obsolete. These RDIs may have become focused on fields of work with limited spillovers for the rest of the economy, and hence have decreased their research intensity.

The trend in Western European RDIs (VTT, SINTEF, Fraunhofer, and Arsenal Research in our sample) is to depend on a mix of income from *unearmarked core funding sources, competitive funding sources, and industry sources.* The unearmarked core funding allows the RDI to invest in generating new capabilities and remain at the forefront of certain technical areas in the long run, while the competitive and industry funding ensure that the RDI remains efficient and accountable. None of the RDIs in the ECA sample display this mix of funding. Funding systems in ECA are still largely homogeneous and have not moved much toward diversified forms of funding: institutional, programs, project funding, and individual grants. Yet, the situation in EU accession countries has changed or is changing significantly when compared with the Commonwealth of Independent States.

Furthermore, core funding in many ECA countries is not distributed in a way that ensures that it is being used efficiently. Most ECA RDIs, including in Turkey, receive their core funding on the basis of their number of employees and operational costs. This is in stark contrast with the situation in high-income countries where core funding is increasingly tied to the outcome of independent evaluations. Independent evaluations ensure that public funding for research and innovation is spent on effective projects, programs, and institutions. In Germany, the individual institutes of the Max Planck Society undergo scientific evaluations by independent Scientific Advisory Boards every two years. At a higher level, the German government requests international commissions to conduct system evaluations of the Max Planck Society. In Russia, RDIs are subject to evaluations, but they are not completely independent and the results do not affect their funding.

Private sector generally lacks a role in the governance of state-owned RDIs

Governance plays a critical role because it influences the overall strategy of the institution, as well as how budgets are allocated and how staff is hired. The composition of the board of directors, also referred to as the "supervisory board" in ECA, provides a good indication of the level of

FIGURE 3.6
Too few private sector board directors

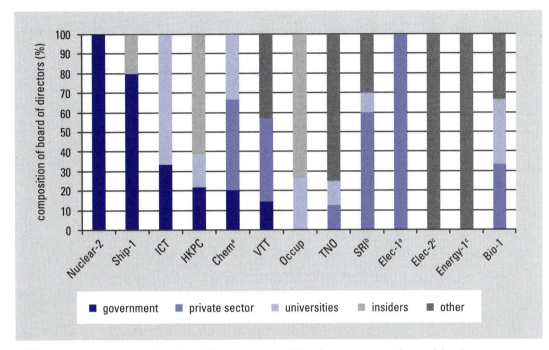

a. RDIs for which the state owns less than a 90 percent share and the private sector owns the remaining share.
b. Private RDIs.
c. RDIs for which the state owns at least a 90 percent of share and the private sector owns the remaining share.
Source: Author calculations based on 2008 World Bank survey of RDIs.

influence of various stakeholders on the governance of the institution. In the ECA RDI sample, only one state-owned RDI has any private sector involvement in its strategic decision-making (Bio-1). In OECD countries, successful public RDIs aim to have a broad range of stakeholders represented on the board. This is the case of VTT in Finland, which, though a public entity, only draws 14 percent of its board from the public sector, the rest consisting of representatives of the private sectors or of various associations. The board of the Dutch public RDI TNO does not include a single government representative (figure 3.6). Including the private sector in the governance of a public RDI allows the RDI to be more closely guided by the needs of the private sector and integrates the RDI in business networks. This provides it with additional commercial opportunities.

Limited autonomy from the government could hinder the performance of some of the RDIs in the sample. In almost half the cases, the state-owned RDIs are not empowered to determine their own budgets. Further, a few extreme cases can be observed where the RDIs are tightly controlled by the state administration. For example, Bio-2 and Bio-3 both require state approval to set salaries and service fees, as well as to set their strategic or business priorities or to offer new services. This lack of flexibility prevents the RDIs from efficiently responding to market needs, including in terms of their research and staffing needs. It is not surprising that both of these institutions rely on government funding for virtually all of their income. Regardless of the level of institutional autonomy of the RDI, in all but one state-owned RDI the executive director is appointed by the state administration. There are no RDIs in which the director is selected by a board of directors with external, nonstate representatives. The major drawback here is that the appointment of the director may be politicized when it is carried out by the government and may result in the establishment of a management structure with insufficient skills and experience. In the United States, the issue of management quality has been addressed by outsourcing the management of some of the national RDIs to private organizations. This is the case of all the RDIs under the responsibility of the U.S. Department of Energy. In some cases, they are managed by private nonprofit RDIs such as the Research Triangle Institute and by private corporations such as Lockheed Martin, and in others by universities. The GOCO approach is not very common outside of the United States.

Weak incentives for the commercialization of knowledge

It is striking that the publishing track record continues to be the most common yardstick to provide incentives for personnel in the state-owned RDIs of the sample, even for those engaged in market-oriented activities. Less than 40 percent of RDIs in the sample provide staff salary-linked and promotion-linked incentives for patenting and generating new contracts. Surprisingly, some RDIs in the sample that derive most their budgets from industry, such as Ship-2, do not provide any incentives to their staff to commercialize their knowledge. Moreover, many ECA RDIs do not have modern human-resource management practices where researchers have clear career paths and salaries based on transparent performance evaluations.

Half the RDIs in the sample employ few or no sales, marketing, or technology transfer staff. Even some of the RDIs that derive important shares of their revenues from the market, such as Ship-1 and Ship-2, have very few commercial staff. This is particularly problematic for RDIs where there are limited incentives for researchers to engage in business development. In three cases, the state-owned RDIs have no dedicated commercial staff. The limited commercial orientation of the management model for state-owned RDIs is further reflected by the fact that several of them have no marketing strategy. A marketing strategy can be seen as a basic first step to commercialize knowledge or at least an "intent" to sell services to the market. All the RDIs in the sample without a marketing strategy are fully owned by the state.

Few of the RDIs in the sample collaborate with external technology commercialization organizations. Only roughly a quarter of the RDIs cooperate with either an incubator or external technology transfer office (TTO). In some countries, such as Russia, the lack of collaboration with TTOs can be partly explained by the legal obstacles prohibiting public RDIs to have shares in spinoffs. This is in sharp contrast with successful RDIs such as VTT and SINTEF that often become shareholders of their spinoffs. Nonetheless, the trend is slowly changing and some ECA RDIs have started to establish TTOs.

RDIs have trouble attracting and retaining a high-quality workforce

The most critical asset of an RDI is its personnel. It is the research staff that accumulates the knowledge that can generate new discoveries and inventions that can be commercialized. Formal education and training provides the fundamental building blocks, but arguably the most valuable knowledge is accumulated by conducting R&D as this results in learning-by-doing and learning-by-interacting with clients and with other providers of knowledge. Universities, particularly in the United States and Western Europe, typically act as national repositories of advanced technical knowledge. For this reason, RDIs in these economies find it useful to forge ties with universities through collaborative projects, graduate student researchers, and faculty. Most RDIs in the sample had ties with universities, by employing students or faculty members. However, the characteristics of the RDI sample confirm the widespread belief that ECA's RDIs are unattractive workplaces for ECA's younger generation of researchers. The average age of the research staff in the sample is high, at times approaching the late 50s, and mostly between 40 and 50.

The aging of the RDI personnel is caused by a number of factors. One factor is that in many ECA countries the number of science and engineering graduates has declined over the years and this is creating supply-side constraints. At the same time, personnel in many of ECA's state-owned RDIs are considered to be civil servants, and inflexible firing rules can make it difficult to replace older researchers. RDIs in Turkey are addressing this issue by hiring an increasing number of project-based staff and placing a freeze on the number of permanent positions. Although this strategy increases the RDI's flexibility, it also makes it more difficult to attract high-quality staff. Moreover, salaries in ECA's state-owned RDIs tend to be significantly lower than in the private sector, as illustrated by the RDIs in the sample (figure 3.7), which make them an unattractive

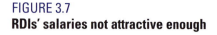

FIGURE 3.7
RDIs' salaries not attractive enough

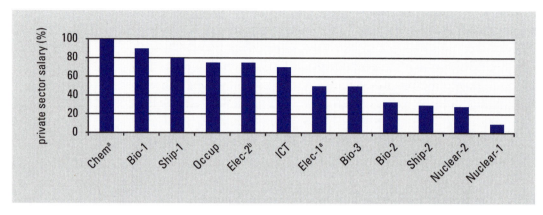

a. RDIs for which the state owns less than a 90 percent share and the private sector owns the remaining share.
b. RDIs for which the state owns at least a 90 percent of share and the private sector owns the remaining share.
Source: Author calculations based on 2008 World Bank survey of RDIs.

option to young graduates. In some ECA RDI such as Bio-2, Ship-2, and Nuclear-2, the RDI salary is less than a third of the private sector salary.

If the performance of an RDI is largely determined by the quality of its workforce, it is worth noting that a number of RDIs in the sample employ few staff with doctoral or even masters degrees. This contrasts with surveys of knowledge-intensive enterprises in ECA countries by Radosevic, Savic, and Woodward (2010), which suggest that these enterprises are able to attract higher shares of skilled personnel with graduate degrees.

A roster of obstacles

In sum, the case studies highlight a number of obstacles hampering industry–research collaboration. A large performance gap exists that needs to be breached for RDIs to fully meet specific technological needs of local firms and complement the national system of innovation by playing a bridge role between pure research-performing organizations and the productive sector. Among the barriers that exist, some can be attributed to *external* factors.

- Mechanisms to finance industry-oriented R&D are underdeveloped. There are few public grant programs to fund collaborative research between RDIs and industry. There are often no business angels or venture capital markets to transfer research coming out of RDIs into early-stage startups.

- In some ECA countries, poor investment climates reduce the demand for innovation in the private sector by making it difficult to establish new innovative enterprises or by increasing the risk of innovation.

- There are legal barriers for collaboration due to unclear intellectual property rights–legislation or inability of state-RDIs to establish spinoff firms.

- There is an increasing shortage of researchers.

- Government funding models based are often biased toward institutional funding instead of competitive funding, thus decreasing competitive pressures to operate efficiently.

- There is little accountability for institutional funding.

A number of *internal* factors relate to the governance of the RDIs.

- Many RDIs do not have the institutional autonomy to operate efficiently.

- The private sector is not involved in the strategic decision-making process of the RDIs.

- There are few institutional mechanisms to facilitate the commercialization of knowledge and knowledge management is therefore not effective.

- Modern human resource management practices are sometimes lacking.

- In some dimensions, like publishing and licensing, the measured outputs of RDIs are very weak.

Many reforms aimed at improving RDI performance in ECA are solely focused on the proximate causes shown in panel 3 of figure 3.8, without consideration for the more fundamental factors included in panel 1, such as governance, funding incentives, and market demand. As a result, reforms have a mostly superficial impact on RDI performance.

For ECA RDIs to effectively integrate their activities in line with private innovative efforts requires the resolution of two interrelated problems. The first is aligning the activities performed by the RDIs to the technological demands of existing business enterprises. The second is concerned with the task of facilitating the flows of information and knowledge at the interface between the RDIs and the community of intended users in the business sector. Their resolution will require addressing both internal and external factors. While some of these factors can be addressed through policy reforms, others will require significant capacity building (figure 3.9). Addressing most of these factors will require overcoming important legacy challenges (box 3.1).

FIGURE 3.8
Chain of events leading to ineffective RDIs

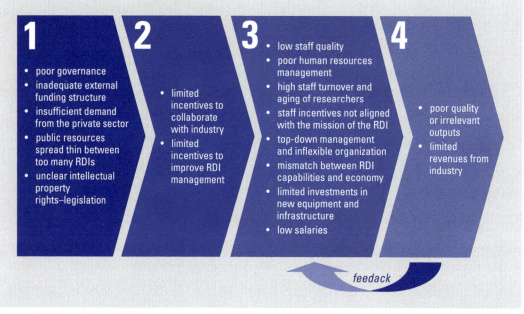

Source: Authors' analysis.

FIGURE 3.9
Factors affecting RDI performance

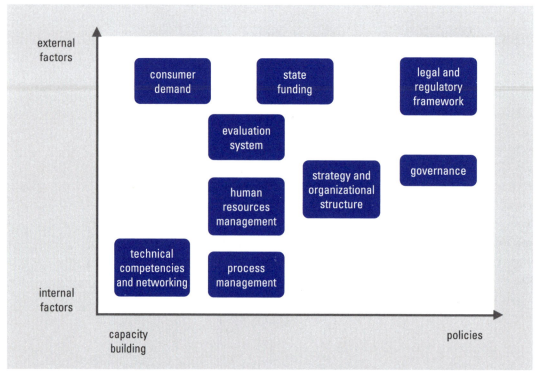

Source: Authors' analysis.

BOX 3.1
Restructuring of RDIs faces important legacy challenges

- *Lack of market experience.* RDIs lived for decades in a protected environment controlled and supported by the state and have a superficial understanding of market-oriented R&D. Their scientists and engineers had limited cooperation with industry. They were working for a single customer, and although some claim their work had potential application, they have little concern about the cost of potential innovation and of market value.

- *Limited experience with application oriented R&D.* They were very strongly focused on scientific achievement, with no real concerns about application. They had a strong interest for basic research, usually quite far from the needs of their local economy. For this reason they are more interested by cooperation with foreign partners, even if they are often not fully competitive at the international level, than to support national industry.

- *"Old fashioned" administration.* Most managers are in their 50s; they were trained during the old Soviet era and are not far from retirement. For these reasons the administration of RDIs is based on outdated principles and faces difficulties to adopt a more pragmatic approach. Most managers are not motivated to stimulate major changes. They have a tendency to assume that the system, even if it slowly degrades and becomes less and less attractive both financially and intellectually, will last at least long enough for them to reach retirement.

- *"Old fashioned" governance.* The management boards have characteristics comparable to the administration. Most members are appointed for scientific and/or political reasons. These boards include a small number of industry representatives. Although they should stimulate the restructuration process it seems that, on a general basis, they are not very much involved with such an issue.

- *A large number of nonproductive employees.* In the past, RDIs had a large number of supporting staff providing nonproductive services (gardeners, cafeteria employees, and the like). Although some scientists left the institution most of the nonproductive employees had very little opportunity to find jobs elsewhere. Consequently the ratio of nonproductive to productive employees, which was often quite high in the 80s often increased again to reach values as high as 25–30 percent of the workforce. As a consequence, the overhead is high.

- *Loss of dynamic scientists.* During the last decade many young and dynamic scientists left the RDIs. They joined some more lucrative position often provided by foreign subsidiaries of western firms or moved to United States or Western Europe where they were offered more financially and intellectually rewarding jobs. As a consequence these institutes lost some of their most dynamic employees, and they have aging staffs who are not motivated to make some important changes.

- *Lack of transparency.* The accounting techniques were not very rigorous, and the financial data were usually very confidential. For these reasons, the figures provided by the institutions during the preparation of the restructuring program are neither very accurate nor very credible.

Source: Nicolaon 2008.

A proposed RDI reform strategy

With so much at stake following the recent global financial crisis and only 10 years to go for the new EU members to attain the EU2020 targets, how can the ECA reform its RDIs in an effective manner? We build on the extensive literature on enterprise restructuring and privatization in ECA on the one hand, and on the evidence from the RDI statistics and case studies on the other hand, to propose strategies for reform of RDIs based on their relevance to national priorities, their expected role as providers of public versus private goods, their performance levels, and their relation to relevant markets and users. When deciding on what ownership and management structure can provide the right incentives for RDIs, governments need to make a distinction between RDIs providing mainly public goods and RDIs already selling or with the potential to sell mainly private goods and services. There is a continuum of possibilities, and any classification of RDIs must take into account that RDIs often produce public and private goods, both at the institution level and within individual teams and projects. Governments also need to distinguish between RDIs whose products and services are developed responding to concrete demands in the market ("market-pull") and those RDIs whose R&D is self-initiated, leveraging a core capability to come up with a technology ("technology-push").[45] This latter dimension is of particular interest to ECA RDIs because it strongly differentiates them from OECD RDIs, which tend to be more demand-driven.

We present below the two components of reform: diagnostic tools to guide RDI reform and options for reform and restructuring RDIs.

Diagnostic tools to guide RDI reform

We classify RDIs in terms of two sets of characteristics, reflecting the market relevance of their activities and the optimal management and ownership structure. Figure 3.10 shows the continuum between private and public goods production on the vertical axis, and on the horizontal axis, between technology-push and market-pull. Against this backdrop, we divide RDIs into groups as a basis for governments to decide which RDI should remain standalone government-owned institutions or merged with universities (quadrant I); government-owned but corporatized, operated by contractors, or organized as autonomous nongovernmental entities (quadrant II); restructured or closed (quadrant III); or privatized to insiders or to outsiders (quadrant IV).

This classification of RDIs according to R&D outputs and organizational characteristics is also a good starting point for discussing the advan-

---

45. See chapter 2 for clarification on the use of these terms in the literature.

FIGURE 3.10
RDI restructuring strategies

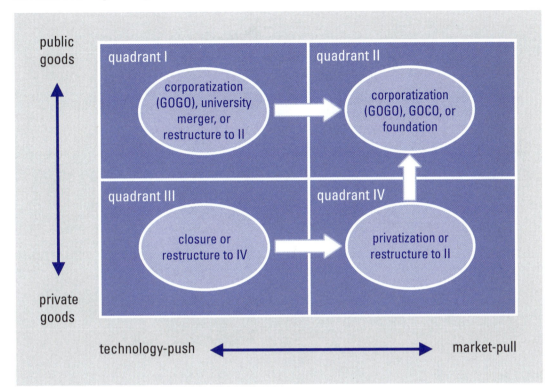

Source: Authors' analysis.

tages of different public funding instruments. If RDIs are producing a large share of public goods (quadrants I and II), then public support through institutional/strategic funding (such as "block grants") is probably needed to partly subsidize recurrent expenditures such as salaries of researchers and strategic assets. RDIs mostly producing private goods (quadrant IV) should not have access to such funding streams. At the same time, competitive funding allocated based on peer review and public procurement can help to top up the budgets of RDIs for high-quality projects in more experimental areas, in the case of RDIs in quadrant IV. This should preferably be through matching grants that provide incentives for collaboration with industry from early on.

Two more characteristics are needed to guide a restructuring strategy, namely the performance of the RDI and the potential market for its services. For RDIs offering public goods, a performance evaluation will identify strengths and weaknesses and help guide reform efforts. High-performing RDIs are not likely to require deep reforms. For RDIs offering private goods, the existence of a market will determine whether to privatize or close the RDI. Figure 3.11 summarizes the questions that ECA policymakers need to address for each RDI.

FIGURE 3.11
RDI reform decision tree

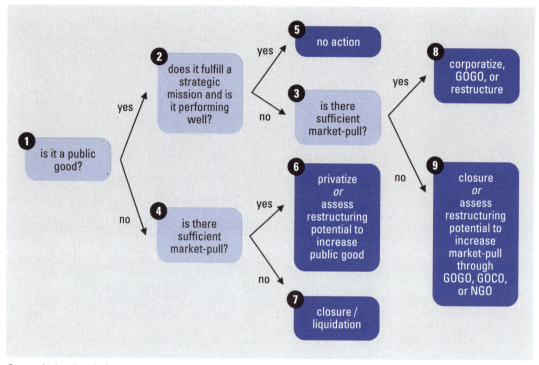

Source: Authors' analysis.

Options for restructuring RDIs

What are the options for restructuring RDIs? There are essentially seven of them (table 3.5).

Option 1: Corporatization and increased autonomy of government RDIs

The government maintains its ownership and management of RDIs but tries to increase their effectiveness by granting them more autonomy. Governments are interested in this option for RDIs producing public goods with strategic implications such as defense, nuclear, and standards that cannot be sustained solely through commercial, current, or prospective clients. Increased autonomy is meant to allow government-owned RDIs more freedom in terms of the direction of the innovation activity, including establishment of small companies. This form of public or state-owned R&D institutes may not be conducive to science–industry collabo-

TABLE 3.5
Restructuring options for ECA RDIs

Option	Relevance to public goods RDIs	Effect on market-pull of RDIs	Effect on RDI governance incentives	Political feasibility
1. Corporatization/autonomy government-owned	+	−	−	+ +
2. Insider restructuring, government-owned	+	±	−	+
3. Government-owned, contractor operated	+	+	+	−
4. Nonprofit foundation	+	−	±	+ +
5. Insider privatization	−	±	−	+
6. Outsider privatization	−	+	+	−
7. Closure/liquidation	−	+	+	− −

Note: These are qualitative assessments about the likely impact and feasibility of the restructuring options based on the interviews with RDIs. For example, the GOCO option is expected have a fairly positive impact in terms of public goods generated, market-pull and governance incentives; however, the political feasibility is fairly weak. The scale goes from very weak impact/ feasibility (− −) to very strong impact/ feasibility (+ +).
Source: Authors' analysis.

ration unless certain preconditions in terms of RDI governance are in place and sweeteners are made available in the form of public subsidies.

Option 2: Insider restructuring of government-owned RDIs

The government restructures the RDIs with the help of its current management by spinning off noncore activities, but maintains its ownership. A restructuring plan specifies which activities will be core activities, which will be integrated to other organizations or "spun off," and which will be liquidated. If an RDI has units conducting basic research, the government can maximize spillovers on education by merging these units with universities. Other options include spinning out activities with high potential for commercialization into an autonomous subsidiary that reflects good practices in competitive funding, governance, institutional design, and internal processes. This "pilot" restructuring program, if successful, can have a demonstration effect on the entire research system and encourage deeper RDI reforms.

Gradual restructuring of R&D institutes requires the full cooperation of management as well as funding and facilitation by a government program to support restructuring of R&D institutes. It is gradual because it is based on bottom-up initiatives by the management of R&D institutes and their financial participation. Ownership after restructuring under this option will remain in the hands of the government while control will remain in the hands of the current management. It will be important to clarify who will be the residual claimant, if there are any profits, which might be the case in RDIs that have stronger links to industry and income from private sources.

Options 3 and 4: Government-owned, contractor-operated or nonprofit foundation

The government contracts out the management of the RDI to an outside contractor but maintains government ownership. The contractor may be a university or university consortium, a for-profit corporation, a not-for-profit organization, or a professional and external management team or CEO. The logic of this option is that government ownership addresses the objective of public good provision and the contractor management facilitates meeting market demand and ensures improvements in internal governance and management. A similar option to GOCO, which requires performance contracts, is to transform the RDI into a nonprofit entity.

GOCO contracts were designed to insulate public organizations from political pressures, and help them to attract and retain talented personnel because they did not have to conform to civil service rules. With management contracts, responsibility for managing, operating, and developing an entity is transferred for a period to a contractor or investor from the private sector who is paid for these services, and simultaneously the level of public funding for operating and investment expenses are agreed upon.

With the GOCO or autonomous foundation approach, governments can develop performance contracts to provide incentives for customer orientation and high technical quality. These can be based on a set of key performance indicators that reflect the overall mission of the RDI and linked to the amount of funding allocated to the RDI. For example, in the case of an RDI that aims to support industrial competitiveness, core government funding could be a function of the RDI's past year performance in transferring knowledge to industry, through research contracts or otherwise.

Options 5, 6, and 7: Insider or outsider privatization and/or closure/liquidation

The privatization and/or closure option is relevant to quadrants III and IV, where private knowledge-based goods and services provide a revenue stream to run an RDI on a purely or mostly commercial basis. Methods of "privatization" for public enterprises include sale through public subscription, sale of shares to employees, or sale to a strategic investor.

We define *insider privatization* as a sale of the company's shares to its managers and workers and *outsider privatization* as a sale to an outside investor—that is, neither a manager nor a company worker. The 1990s privatization of RDIs in ECA is considered to have had mixed outcomes. In several countries, the privatization led to acquisitions by investors interested in the valuable real estate possessed by the centrally located RDIs. The investors then typically disbanded the RDI and used the real estate to develop shopping malls and for other commercial urban uses. One way to deal with the concern about assets is insider privatization.

If the government as owner of an RDI concludes that none of the other options in table 3.5 can resolve the RDI's problems, the last resort is closure. Obviously, politically this is the most difficult option, and we show it in the table with a "double minus." To the extent that the RDI produces public goods, the effect of closure on the economy is negative as it can erode research capacity in a specific technological area. But to the extent that the RDI sells private goods, the effect of closure is positive: it limits "crowding out" and levels the playing field vis-à-vis private companies (usually SMEs) that deliver the same products and services.

Options for public funding to support RDI reforms

Governments can also create incentives for RDIs to become more market-oriented through their research-funding strategies. The first strategy is to restrict the amount of unconditional core funding allocated to the RDI (figure 3.12), so that the RDI must seek revenue from the market to remain financially sustainable. Another strategy is to introduce *earmarked* core funding. What this means is that government funding is still granted unconditionally but is allocated ex ante to specific activities, equipment, or infrastructure within the RDI, which ensure that the RDIs activities are relevant to private sector needs. Although this provides the government

FIGURE 3.12
The less the government funding, the more market-pull dominates

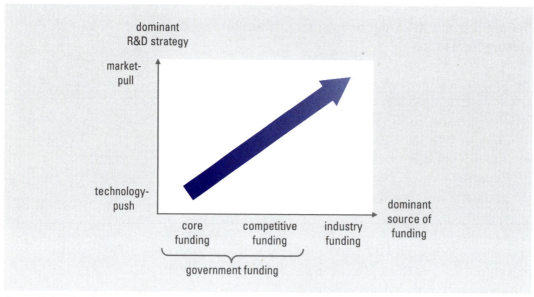

Source: Authors' analysis.

with closer oversight of the RDI's activities, it further centralizes research funding decision-making by displacing the resource allocation challenge from the RDI's management to the government's bureaucratic apparatus. This approach is only useful if the government's resource allocation strategy genuinely reflects the country's industrial needs. However, this strategy does not provide incentives for increasing the institution's efficiency.

Another public funding strategy that can address research quality and research strategy issues is competitive (or "grant") project funding. In this case, funding is allocated on a competitive basis to research groups for particular projects and for particular outcomes. In theory, this approach ensures that funding is allocated on the basis of research quality and efficiency and introduces accountability, given that unsatisfactory outcomes for one grant will hinder a research group's chances of obtaining a second one. In OECD countries, competitive funding typically constitutes the largest share of government funding. Governments have focused on decreasing the ratio of institutional funding to competitive project-based funding to allow for better monitoring of results and accountability of public sources of funding (OECD 2003). A survey of large European research institutions showed that they tended to receive roughly a third of their funding as core funding, another third as competitive public funding, and another from contract research (EARTO 2005).

However, competitive funding has its limitations as well, particularly in countries with limited research infrastructure. In such countries, there is often a single predominant RDI focusing on a particular technical field, which implies there will be little or no real competition for government grant funding. Given the critical mass of equipment and researchers, and the lumpiness and learning curve of research, it is going to be nearly impossible for new entrant RDIs to displace incumbents or for existing RDIs to encroach on each others' technical areas unless there is a provision for international consortia that can leverage the strengths of domestic and foreign partners to compete for these funds. The latter is not always politically feasible in a context of limited resources, as this involves a transfer of subsidies abroad. Another limitation has to do with the peer-review process that allocates grant funding, as this takes time to develop and must be designed to operate in a transparent and independent way for grants to go to the best projects. The result is that competitive funding in certain countries often has a similar effect as direct earmarked subsidies.

Case study: Finland's shift to a knowledge-based economy: The Role of TEKES

Over the past 20 years, Finland has transformed into one of the most productive and innovative economies in the world—managing to narrow the labor productivity gap with advanced Organisation for Economic Cooperation and Development (OECD) countries, at a time when Europe as a whole has struggled to increase labor productivity. In 1990, the GDP per hour worked in Finland was 75 percent of that in the upper half of OECD countries by income, but by 2008, it had increased to 85 percent. Finland's aggressive innovation and the development of a dynamic information and communication technology–industry was a key driver of this performance. Finland's experience is highly relevant to many ECA countries, given their similarities as small, open economies positioned at the geographical edge of the European Union.

Finland's case epitomizes the successful use of public policy instruments in promoting innovation. The levels of research and development (R&D) funding increased by a factor of 5 since the 1990s. Its current R&D/GDP ratio is 3.5 percent, the second highest in the world. Remarkably, much of the growth in R&D spending has been driven by business R&D, with the share of public funding in total R&D remaining below the OECD average. The increase in public funding has stimulated rather than crowded out private R&D. The process leading to this research focus commenced in the 1980s, when Finland established the National Technology Agency (Tekes) to distribute funding for R&D, and augmented its functions by creating the Science and Technology Policy Council in 1987. One of the achievements of the targeted innovation funding has been a significant increase in the R&D activities of the private sector. Hours spent on R&D in the business sector increased by 112 percent from 1991 to 2007 (figure 3.13).

FIGURE 3.13
Finland's business sector is sharply stepping up its R&D

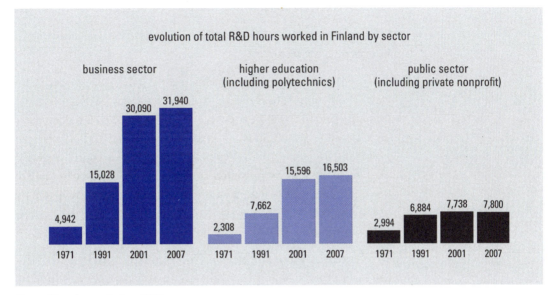

Source: Veugelers and others 2009.

In 2009, the Finnish government supported R&D, including basic research, for a total of €1.9 billion. The public R&D support is highly geared toward support of applied research and commercialization, with 30 percent of the public funds—representing the highest individual share of the budget—being targeted at the support of commercially oriented applied R&D through National Technology Agency of Finland (Tekes), and a further 16 percent targeting applied research through government research institutes. Universities receive 26 percent of the funding; the Academy of Finland receives 16 percent and other institutions 12 percent.

Finland's highly integrated and dynamic national innovation system provides the backbone of the country's innovation success. Figure 3.14 illustrates the various channels and institutions that provide funding for R&D innovation at various levels of the innovation chain.

- The Finnish National Fund for Research and Development (SITRA), is a publicly owned venture capital fund that is managed independently, but operates under the responsibility of the Finnish Parliament. SITRA also operates funding schemes for pre-seed startups and acts as information broker and network platform for pre-seed innovators through its LIKSA, INTRO, and DIILI programs.

- The Technical Research Centre of Finland (VTT) is a public research institution focused on applied R&D in a variety of disciplines from building technology to electronics. While being publicly funded, VTT has a strong track record of providing R&D for private sector firms, which account for roughly one-third of its revenue.

- Support through the Finnish Academy of Science and universities focuses on basic research.

- At the heart of the Finnish policy framework stands Tekes, which provides grants, and subsidized and convertible loans ("capital loans"), primarily geared toward early stage technological development. Having a broad and ambitious mandate to look ahead and identify the promising areas for technological advance TEKES coordinates the working of the innovation system with the help of catalytic funding of R&D. It works closely with government agencies, with the Academy of Finland which promotes basic research, with SITRA the Finnish National Fund for R&D and with universities.

In recent years, most of the support funds administered by Tekes were in the form of neutral grants, with close to half the agency's funding going to technology programs. These programs have ranged from public

FIGURE 3.14
Finland's successful innovation environment

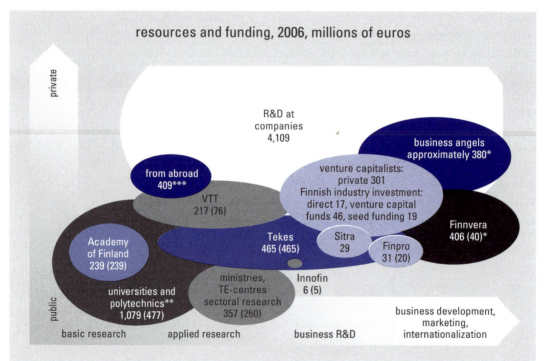

Note: The figures represent the total extent of each organization in million euros in 2006, those marked with star are earlier. In parenthesis the share that is funded from the State budget. **includes polytechnics and *** includes R&D costs of corporations foreign units.
Source: www.tekes.fi/en/document/42610/rd-finland_ppt accessed on November 4, 2010.

health-care technologies to nanotechnology and business and management technology.

The economic rationale of the technology programs is to enhance R&D cooperation between different companies, public R&D institutes, and international actors, along with transferring knowledge and skills among the participating entities. Internal evaluations of initial programs find positive returns to promoting R&D cooperation and coordination. But the success of the programs relies equally heavily on the quality of public administration in identifying and deciding on relevant program areas. Information feedback mechanisms and coordination with local R&D institutions and industry are particularly important in formulating the technology strategy.

The cooperative model of public policy formulation in Finland fosters a bottom-up approach to developing the technology program priorities, thus avoiding some of the risks of top-down industrial policies. Interest and active collaboration exists at the highest level to support innovation. Research and innovation policies are taken with the input of several ministries, including the Finance Ministry, Ministry of Education, and Ministry of Employment and the Economy. A recent evaluation of Tekes and the National Innovation System looked at the participation of top government officials in the Research and Innovation Council. Results showed that the prime minister attended the council 97 percent of the time. The ministers of Economic Affairs, Education, and Agriculture and Forestry also had high attendance rates.

Can ECA emulate the Tekes' model? The answer is not likely. In Tekes, the decision-making process, project selection, and formulation of programmatic priorities are heavily dependent on the quality and capacity of public servants, as well as the Finnish governance model with the virtual absence of corruption and capture and a transparent and cooperative approach to public policy formulation. The absence of most of the latter implies that the Tekes model cannot be transferred as is to many ECA countries. Instead, ECA governments should focus on adopting the funding instruments but complement the decision-making processes with independent external control and oversight through peer reviews and foreign experts. As for programmatic priorities, ECA governments should be encouraged to emphasize neutrality in the early stages of the development of funding programs and focus on technology policies based only on ex post patterns occurring over time.

Chapter 4 will deal with the financial instruments that governments can use to foster commercial innovation—including the tradeoffs that exist between neutral and targeted interventions, and between instruments that are aiming to develop a pipeline at the very early stage of technological development and those seeking to provide access to finance to fast growing startups through venture capital and other mechanisms.

CHAPTER 4

Bringing innovations to market—boosting private incentives through public instruments

▶ To bring innovation to markets, governments need support instruments that promote private risk-taking and stimulate markets for private risk capital.

▶ Some support instruments have proved more effective than others in OECD countries, but it is also abundantly clear from the experiences in emerging countries that they will not work without a friendly investment climate, adequate intellectual property legislation, and built-in mechanisms to avoid corruption and regulatory capture in project selection (such as international peer review).

▶ Financing instruments need to be designed so they avoid crowding out private investment and other funding sources (by requiring the matching of public funds with private cash contributions—"additionality")—and minimizing distortions ("neutrality").

▶ **In an environment where absorptive capacity is weak or where there is little tradition of commercializing scientific results, the government needs to carefully sequence its provision of support for innovation—ensuring this is provided throughout the entire commercialization cycle, but with different policy instruments in the three main stages: early stage, growth, and exit.**

▶ **Effective sequencing should aim at building a significant pipeline of early stage projects before supporting venture capital. If not, the venture capital will likely fail as it will lack the critical mass of projects to create a good portfolio. The success of the growth stage depends on a deal flow of attractive companies coming out of the early stage.**

Governments can support bringing innovations and technology to market in a variety of ways. At the most basic level, effective government policies should create an institutional base that establishes openness to trade, improves the business environment for investment (including foreign direct investment, FDI), establishes effective intellectual property rights regimes, and enhances knowledge flows—by improving the ability of academic and research institutions to generate the specific research and development (R&D) projects that attract private investment by firms and investors. Beyond those general policies, most governments have also intervened at the firm level to stimulate private funding of R&D on the basis of the arguments of market failures and the capital gap for funding innovative technology-oriented firms.

The government needs to carefully design the instruments to promote private risk-taking rather than rent-seeking and stimulate markets for private venture capital. It should not decide ex ante which technological sectors (with the exception of those linked to public goods), firms, or projects to support, but rather should respond to market demands. Especially in Europe and Central Asia (ECA) countries, the institutional design should aim to immunize, as much as possible, the funding allocation from interference by political actors, corruption, and capture by other state or specific interests. And the government should provide support for innovation throughout the entire commercialization cycle—drawing on different policy instruments for different stages of enterprise formation:

- *Early stage.* Incubators, angel investors, or matching grants, as well as spinoffs and other spillovers from multinational corporations.

- *Growth stage.* Government support for private venture capital through risk-sharing.

- *Mature stage.* Facilitating access to international and local equity funds and entry by strategic investors.

One of the major challenges ahead for policymakers is the sequencing of support for early stage vis-à-vis support for growth stage innovation, mainly venture capital. Effective sequencing should aim at building a significant deal flow of early stage projects before supporting venture capital. The success of the growth stage depends on a deal flow of attractive companies coming out of the early stage.

So what type of interventions should the ECA countries adopt? In this chapter, we begin with a look at the basic principles of instrument design and how they can be applied to ECA countries. Next, we review the basic type of financial instruments used in Organisation for Economic Cooperation and Development (OECD) countries—including grants, loans, tax incentives, and procurement preferences—and examine their applicability to ECA. Then we explore the recommended financial instruments for the ECA region—(such as R&D matching grants, loans, venture capital, and tax holidays)—along with institutional instruments (such as R&D institutes, incubators, technology parks, and technology transfer offices).

It is important to differentiate between financial support instruments, which is the main focus of this chapter, and nonfinancial support instruments. Both are government subsidies to private entrepreneurs, but, though the nonfinancial instruments combine subsidization with the public provision of the subsidized service (such as a government-owned and government-run incubator), a financial subsidy allows the entrepreneur to spend the subsidy on buying the business services (for example, from the incubator) or to invest the subsidy in equipment (such as a prototype) or working capital. That said, both types of instruments are complementary, reinforcing each other.

Basic principles of instrument design

OECD countries have been experimenting for decades with several instruments to support commercial innovation—and many ECA countries, especially the new European Union (EU) member states and accession candidates, already operate variations of these schemes. What can the OECD teach in terms of effective program design? Three principles emerge. First, it is essential to evaluate the *institutional environment.* Given the institutional and governance situation and the identification of corruption as one of the main constraints to the business environment in many ECA countries, it is of the utmost importance to protect projects from misappropriation by the state. Second, it is vital to provide *additionality*. Any instrument needs to aim at avoiding crowding out while promoting private investment and risk sharing. Third, it is critical to pursue *neutrality*. To minimize distortions, governments should avoid sector and company targeting ("picking winners"), with the exception of sectors associated with public goods such as health, the environment, or security.

Institutional environment

The design of new instruments needs to account for the existing institutional environment, with an emphasis on weighing the benefits and potential for effective restructuring of existing instruments against the advantages of creating new institutions and instruments from scratch. An illustrative example of the use of effective financial instruments supported by a conducive institutional environment having fostered private sector innovation is Turkey's renewable energy sector (box 4.1).

To avoid government capture and failure, instruments should be designed to be as neutral and transparent as possible. Most critically, the decision-making (selection) processes about funding allocations need to ensure that the quality of selection is driven by true innovative and commercial potential. The continued presence in many ECA countries of corruption and capture of governmental processes by interest groups places a heavy burden on the design of successful policy instruments. The various grants and venture capital funding proposed under the project are likely to attract rent-seeking behavior, which could result in inefficient funding allocations if the institutional design cannot immunize the funding allocation from interference by political actors and other interest groups.

The design of instruments is crucially dependent on the capacity of public servants to administer them and insulate their decision-making promises from capture and rent seeking. As the Finnish case study shows, some of the most successful innovation support systems in the world rely heavily on the analytical and managerial skills of public servants to take good economic decisions (see chapter 3). Although Finland is successful with this setup, it is questionable whether the model can be implemented as such in many ECA countries. Weak public service institutions might result in a lack of capacity to make informed and economically beneficial decisions.

Instrument design in ECA thus needs to enhance the decision-making processes with sufficient checks and balances through a wide representation of private sector, academia, civil society, and foreign expertise to

BOX 4.1
Catalyzing private sector innovation in Turkey through an improved institutional environment and financial instruments

Over the past decade, Turkey has been overhauling its energy policies to meet growing energy demands, ensure an economic and efficient energy supply, decrease fossil fuel use, and reduce greenhouse gas emissions. Its recent efforts in the renewable energy market stand out as an example of how a better institutional environment coupled with appropriate financing instruments can lead to private sector innovation with powerful economic, social, and environmental benefits.

The key changes began in 2001, when the government launched new energy strategies and enacted energy legislation aimed at liberalizing the electricity market, promoting privatization, introducing competition, and establishing an independent regulatory regime. As part of this effort, it prepared *The Renewable Energy Project*, with the World Bank's support, to "expand privately owned and operated distributed power generation from renewable sources without the need for Government guarantees and within the market-based legal framework of the new Electricity Market Law." The project supported institutional development activities for introducing laws—such as the 2005 Renewable Energy Law, aimed at bringing Turkey more in line with European Union regulations—mechanisms, and procedures for private investment in renewable energy. It also funded a Special Purpose Debt Facility, which provided a $201 million long-term lending facility for renewable generation. Financial intermediaries used the facility to provide long-term debt financing to private sponsors of renewable energy subprojects.

One of the most important subprojects was the Mamak Landfill Gas Power subproject, implemented by ITC Mamak, a private company, in 2006. It is one of the first large-scale landfill-gas use projects in Turkey—and one that has demonstrated a new approach to converting waste into energy in Ankara's metropolitan region. Investments in research and development were critical for the project's success and the company's continued growth. Stimulated by a more business-friendly environment for renewable energy innovation and financial incentives, ITC Mamak developed a way to create energy from mixed waste that cannot be treated normally through integrated waste management systems. To achieve this, it collaborated with foreign research institutes, foreign companies, and equipment suppliers to not only implement existing technologies but also develop and patent new technologies in Turkey and abroad. These technologies contributed to novel concepts for the company's sorting plant, waste fermentation system, and gasification unit. Following the project's success, similar projects are now being planned and constructed throughout the country. The government envisions that 5 percent of the country's energy will be produced from garbage in the future.

Besides energy generation from potentially toxic waste, the subproject has markedly improved the lives of people living around the site. For more than 20 years, the waste of 4 million people living in Ankara was stored in an uncontrolled manner. The residues caused environmental and social problems including pollution (and thus climate change), strong odors, health risks, and even potential explosions. As a result of the Mamak Landfill Gas Power subproject, the risk of gas explosions is reduced, seepage of explosive methane gas is decreased, local air and water quality is improved, a Waste Water Treatment System that improves soil condition has been installed, and 4,500 trees have been planted around the landfill area.

Sources: World Bank 2010b; firm interview with ITC Mamak Project CEO.

protect the decision-making process from rent-seeking behavior and cap-ture by interest groups. An optimal instrument design should include the following key elements:

- The administration and funding decisions are located in an indepen-dent institution with a clear mandate and control mechanism, separat-ing it from other public policy goals.

- The funding decision is made by an independent investment commit-tee. To enhance transparency, it is advisable to staff the investment committee with technical experts and foreign experts that are less likely to be subject to political influence. A potential problem is the question of confidentiality and fear of industrial espionage.

- The investment policy and decision processes are instituted and super-vised by a supervisory board consisting of representatives of different government institutions and international advisors.

- Technical assessments of the project proposals are based on external peer reviews involving international experts where possible.

All project proposals and decisions are recorded, tracked, and made pub-licly available to enhance transparency. E-government procurement technologies should be considered to aid the process.

Additionality or crowding out?

Another key design question is whether government support programs create new investment in R&D or simply crowd out private investment, which is substituted by government funding. Most assessments of these types of programs in OECD countries are based on aggregated statistics such as the volume of financial flows, as well as anecdotal and intuitive evaluation of the relationships between policies and the subsequent eco-nomic performance of an economic sector. Empirical evaluations using counterfactual datasets—that is, what would happen in the absence of the intervention—are few.

So far, the evidence is mixed. In a study of the U.S. government's long-running Small Business Innovation Research (SBIR) R&D grant program—which compares program awardees and a matched sample of firms that did not receive awards during a 10-year postwar period—Lerner (1999) found that firms receiving grants grow significantly faster than the others after receipt of the grant. But his results are ambiguous in suggesting that the effect may relate more to "quality certification" by the government, enabling the firm to raise funds from private sources. Indeed, his findings suggest distortions in the award process; companies receiving multiple grants showed no increase in performance. Wallsten (2000) found that the SBIR program crowds out the firm's own research spending about dollar-for-dollar, reversing the finding of Lerner (1999) for this same program. Trajtenberg's (2001) review of a number of studies of Israel's R&D grant programs suggested that there is evidence, though limited, of a positive relationship between the grant programs and productivity in R&D-intensive industries. Lach (2002) found that research support of commercial firms in Israel increased the firms' total R&D expenditure by $1.41 for every dollar of public research expenditure. Branstetter and Sakakibara (2002) found that Japanese funding of research consortia increased the R&D of the participating firms. And Ali-Yrkkö (2004 and 2005) showed that the increase of public funding in Finland did not lead to a crowding out of private R&D funding.

Thus, government interventions need to be carefully designed so that they do not crowd out private investment and funding sources. Although financial market failures can be identified, especially in the early stages of innovation, the smaller the distance of the innovative process from the market and the higher the probability of market success, the higher the probability of financing from mainstream financial intermediaries. It can be argued that the important principle of matching may prevent or at least mitigate crowding out. Projects closer to commercial application should be funded by venture capital or other private sources.

As much as possible interventions should be designed to promote private risk-taking and stimulate the private risk capital market. A number of design issues can be taken into account.

Risk-sharing. The high uncertainty about technological and commercial success in the early stage technological developing (ESTD) phase not only deters mainstream financial institutions but also represents a risk for the innovator. Often, the inherent uncertainty of success is the key obstacle in providing incentives to potential entrepreneurs to invest their own money, accommodate the opportunity costs of leaving a secure job, and take commercial risks by borrowing money.

Preservation of incentives. The design of the instruments needs to preserve the incentives for entrepreneurs to invest their intellectual resources and time and effort in the pursuit of success. Concessionary funding is prone to "moral hazard" problems.

Commercial orientation. Criteria for funding decisions need to clearly distinguish between projects that are technologically interesting and the targeted group of projects that are technologically innovative and have potential for commercial success. Commercial success potential must be a criterion for project selection.

Specific bottlenecks. The choice of instrument varies according to the different stages of the innovation chain. In some ECA countries, the most effective set of interventions will be combinations of financing instruments and measures to enhance innovative capacity and reforms to the business environment. The optimal level and degree of subsidy should be lower, the closer the intervention target is to functioning market mechanisms.

Neutrality

Neutrality of government programs supporting innovation (such as matching grants) means that the government does not decide ex ante which technological areas, firms, or projects to support, but rather responds to the demands coming from the market. Under that approach, the government sets universal criteria for submission and eligibility (for example, technological and commercial viability, proven business record). The entrepreneurs—that is, the would-be innovators—submit project proposals for support, and the agency in charge supports those that best fit the criteria.

More generally, neutrality means that the program should not try to steer the grants (or other instruments) in any predetermined direction but rather should try to deploy them in such a way as to maximize spillovers or social returns. The success of R&D support programs in Finland and Israel is in large measure attributed to the fact that the policies implemented were largely neutral in that sense. There were still instances of targeting, but the thrust of the policies remains neutral. Today, Finland has established specific sector programs; however, the emergence and selection of these specific sector programs are driven by an ex post recognition of clusters that have emerged in a neutral and competitive policy environment.

The main critique of the merits of neutrality in this type of intervention is that, in the first place, the rationale given above for intervention is the presence of spillovers—that is, a gap between the private and social rates of return. The difference between the social and private rates of return may be more than a factor of 3 to 4 (Jones and Williams 1998). Yet, this gap may not be constant across projects. Suppose we have two projects with identical private rates of return, but one has a social rate of return marginally higher than the private return, and the other has a social rate of return higher by a factor of 10 than the private rate of return. The market will be indifferent between the two, though from a social perspective, the one that offers a high social rate of return is preferable. If all this information is available, neutrality is not the best policy.

Some countries in East Asia have taken a mixed approach, adopting project and firm neutrality while targeting certain sectors or industries—known as "industrial policy" (see chapter 1). However, selecting projects based on "social return" is in most countries impractical because it requires huge amounts of information about the parameters of each project's social benefit. Moreover, because such parameters are necessarily subjective, selection based on the social benefit of each project opens a "Pandora's box" of capture and corruption possibilities. Given limited institutional capacity, it is unlikely that many ECA governments would be able to estimate the social rates of return and select projects with the highest social benefit.

Thus, we should opt for neutrality as a second-best option—given the threat of corruption and capture by vested interests (for example, old firms or organizations for whose services there is no demand but who are trying to survive with state aid for innovation). Only after a track record of excellence has been established, with several years of experience (as has happened in Finland and Israel) should the principle of neutrality be modified—but only toward sectors or industries, not individual firms or projects. Moreover, neutrality does not imply lack of choice criteria: the authority in charge of administering the program will choose among competing projects, using certain criteria. The question is what criteria guide the choices and what the information requirements of such criteria are. (For example, the authority administering the program could choose those projects that maximize spillovers, though information requirements may make this criterion difficult to implement and open the door for unwanted influence.)

Basic types of instruments

Grants versus loans

What are the two most basic types of instruments for government intervention to promote investments in innovation and R&D? The first is direct subsidies—one form of which is grants, which typically require some share of matching investments by the grant recipients. Grants have two clear advantages over loans for the promotion of innovation. They provide funds through matching grants to reduce the entrepreneur's risks, which is typically the most important constraint in providing incentives to innovators to pursue commercial applications. In the case of technological or commercial failure, the loss to entrepreneurs is limited to their own matching investment, and they do not have to pay back the grant.

In addition, R&D and innovation activities require high up-front investments that may generate positive cash flow of an uncertain level at some future point. Grant instruments (like the equity provided by private risk capital investors) can support this investment profile by providing the needed up-front investment without crippling the company or project with mandatory payments before the positive cash flow can support them. Many of the most successful grant programs are designed to mimic the positive cash flows with royalty payments on successful programs.

The second type of instrument is those with mandatory repayment (*loans*). This instrument, such as commercial loans or even loans with interest rate subsidies, provides neither the crucial risk-reducing feature nor the cushion of support for the cash flow. Mandatory repayments may starve a potentially successful project of internal financing to invest in later stages of development and commercialization. In the case of technological or commercial failure, entrepreneurs not only lose their own investment but also have to repay the loan in full. And an entrepreneur is unlikely to consider engaging in already risky innovative activities when the risk is compounded by the high prospect of business failure from loan foreclosure, potentially leading to bankruptcy.

The mandatory payment structure of loans makes them unsuitable instruments for ESTD projects with uncertain cash flows and unknown ex ante prospects of success. Even so, loans can play a role at later stages of the innovation process in which the risk to the entrepreneur declines with a greater probability of success and reduced distance to market. When looking at best-practice examples implemented globally, it is important to review the economics of the instrument employed. Various instruments can be found that are nominally classified as loans—however, they contain provisions that forgive the loan repayment in case of project failure or convert the loan into an equity participation, thereby reducing the ex ante risks and disincentives to the entrepreneur and effectively assuming the economic advantages of a grant mechanism combined with a "tax on success." Box 4.2 provides an example of a combination of matching grants and conditional loans instruments in Croatia.

BOX 4.2
Using grants and loans for innovation support in Croatia

For firms wanting to conduct research and development (R&D) in Croatia, scarce public support for R&D is a main binding constraint. Croatia's public R&D spending is about 0.7 percent of total government expenditures, while countries such as Spain have increased this share from 1.5 percent to about 2.4 percent in the past 10 years. In many ECA countries the situation is similar to Croatia's in that the returns expected from innovation and R&D are high, but investments by the public sector are low.

To address these challenges, the Croatian government asked the World Bank to provide support in developing and implementing its new and comprehensive science and technology policy. The Science & Technology Project was conceived as a backbone of this strategy. With a budget of €31 million and a five-year time horizon (2005–11), the project had two main objectives: enhance public financing for business R&D, and foster the commercialization of public R&D. The Project employed numerous funding instruments to cover the different stages of the innovation value chain, including knowledge creation, transfer, and use.

One of the Project's main components is the SPREAD program, which supports knowledge transfer and R&D commercialization. It tries to foster close links between industry and R&D institutions, encourage industrial firms to substantially increase R&D activities, use the existing infrastructure in technology institutions to the fullest extent possible, and help industrial firms make R&D projects more cost-effective. It funds up to 50 percent of the cost of joint projects between small and medium enterprises and R&D institutes through nonreimbursable matching grants, with a maximum contribution of €120,000 (originally €80,000). By the end of 2010, 17 projects were signed for a total of €1.7 million, and 5 more were expected to be added in early 2011. Most projects are with micro and small companies, chiefly in the information and communication technology and electronics sectors.

Another key Project component is the RAZUM program, which supports knowledge use and innovation. It employs a soft loan mechanism to encourage the private sector to spend more on R&D by sharing the risks, particularly in the later stages of product development. The conditional loans cover up to 70 percent of new product development costs with a maximum support of €1.5 million per project and a duration of up to three years. Loan repayment is conditional upon the project's success, with repayment 3–5 percent of sales. As of March 2010, the program supported 16 projects with a total of €11.5 million, €7.5 million of which came from the European Structural Funds and the rest from the Croatian government. The private sector contributed €9.3 million.

What is the verdict so far? Preliminary evidence suggests that SPREAD has helped commercialize public research. Beneficiaries indicate that without grants, they would not have collaborated with these organizations or developed their projects (at least with the same speed and quality). A success story is Inteco, which developed an automated robot for hydrodynamic treatment of iron and concrete in construction in collaboration with Zagreb University. As for RAZUM, early results suggest that without the program's support, firms would have lacked sufficient financial resources to carry out the projects or suffered delays on the way. Another success story is BANKO, a family company from Split, which designed and constructed a sliding vane pneumatic grinder with superior process quality, enhanced durability, and reduced energy consumption. In June 2010 BANKO concluded about 150 contracts with customers in Germany, the Netherlands, the Russian Federation, and Ukraine.

The bottom line is that the strong interest in these programs confirms a high demand for programs that facilitate private investment in R&D, and that innovation activity is constrained by limited access to financing—not by a lack of ideas or projects.

Sources: Correa, Tarade, and Borowik 2010; company interviews.

Tax incentives and procurement preferences

How about other instruments that have been used internationally to promote investment, innovation, and R&D? Rather than directly allocate government funding for projects, these instruments attempt to indirectly stimulate investments through procurement preferences and tax incentives. Because they act indirectly, however, they must respond efficiently to the preferences and incentives of public and private actors. Thus, they are a more complex policy lever, and in ECA, with weaker institutional contexts, they are likely to be tough to implement effectively.

Tax incentives have been widely used, especially in Europe, to encourage R&D investment across a wide range of firms. Tax incentives encourage investors and companies to invest in R&D or new companies through tax credits and lower tax rates. Tax incentives can come in several forms: up-front tax credits for investments, reductions in capital gains taxes or tax rates on investments, and tax credits to offset losses from investments in small and medium enterprises (SMEs) or equity funds. Tax incentives are in principle neutral—they apply to all qualifying firms equally and thus uphold one of the key principles in instrument design. Moreover, in the United States, a study by Hall and van Reenen (2000) suggests that a dollar in tax credit for R&D stimulates a dollar in additional R&D.

That said, tax incentives have a number of weaknesses that make them less applicable to the ECA region.

- First, tax benefits would help existing enterprises that can use profits from related products to take advantage of the credits or offsets. But tax incentives do not help startups that have not yet accumulated sufficient profits and thus cannot offset tax liabilities. Innovative startups may have very low profits for a long time—meaning that tax benefits would provide no funding at the critical initial period when commercialization takes place.

- Second, in countries with a weak tax enforcement system, tax incentives may promote distorting tax avoidance behavior rather than productive investment. There is a risk that companies will reclassify expenditures without justification by presenting regular expenses as R&D expenditures. Coping with such tax avoidance or evasion requires a highly sophisticated tax inspectorate, which is unavailable in most ECA countries. Lack of specificity and poor design of the tax code can also limit the impact of tax incentives because some firms are able to benefit from reduced taxes without having significantly altered their behavior.

- Third, because tax incentives are indirect, the fiscal cost of support is not fully apparent in the budget and thus hidden from policymakers, while financial support in direct programs is easily observable in the budget.

- Fourth, tax incentives cannot be used like grants to promote the creation of networks and links between the private sector and universities and research institutes, which lie at the heart of this type of instrument.

Another indirect instrument, procurement preferences, are a variation on a direct grant program in which a portion of existing government research budgets are earmarked for small innovative firms, as in the SBIR Program in the United States (see chapter 3, box 3.2). This type of program is most effective in large economies with significant government-sponsored, commercially oriented research budgets and transparent procurement processes—conditions that are unlikely to exist in most ECA countries.

Financial instruments for ECA

What types of financial instruments would be most useful for ECA countries, where the funding gap—what federal and state governments must provide to make up what is not funded by angel investors or venture capital—is so much larger than in industrial countries? To answer this, it is important to recognize the underlying reasons for the funding gap, which are connected to the low capital accumulation before 1990 and the shortage of entrepreneurial capacity in corporations. As figure 1.3 showed (see chapter 1), about 60 percent of ESTD funding in the United States comes from angel investors or corporate venture (internal financing by corporations), and 34 percent comes from federal and state governments (such as SBIR). In the ECA region, however, internal financing by enterprises and angel investors is rare in the region and does not provide a viable basis for promoting innovation. In this context, "angel investors" refers to successful entrepreneurs that look for new opportunities to invest private funds (earned from their own previous innovations) and are willing to invest in ESTD projects in technological fields that they understand well (having "been there and done that").

It is vital that state intervention in this environment deal with the bottlenecks in all stages of the innovation chain from the generation of ideas to commercialization. And here it is important to stress that the supply of ideas (the pipeline of projects) is endogenous: potential inventors and entrepreneurs need to be confident that if they create projects there will be funding available for them. Why else would they invest resources in innovation? Therefore, funding programs cannot wait until ideas appear by themselves, but rather all stages of the innovation chain need to be dealt with in parallel.

Taking these design principles into account, there are three types of instruments that stand out as the most useful for the region: minigrants, matching grants, and early stage venture capital support. They address weaknesses at different segments of the innovation value chain in the ECA region, with limited distortions and clear objectives. Minigrants provide small grants directly to support precompetitive R&D, and matching grants require companies to match the investment. Government leverage of private venture capital support promotes the risk capital market, which will eventually become the long-term driver of innovation investment.

Business support services are necessary to develop projects that would be acceptable for funding. It is important that they be viewed not as a standalone policy lever but as a complementary support to the core instruments that provide financing for innovation and R&D. The design of such a program will also have to play a coordinating role in the number and types of instruments that a country puts in place to support commercial innovation. There is a certain sequencing of instruments that matches the innovation process—researchers and entrepreneurs have to be aware that these instruments are available for the entire innovation process and that the support system has been designed in a comprehensive and coordinated manner. Not only will that ensure the availability of support for the entire process but it will also help curtail inefficient duplication of efforts by several agencies.

Minigrants

Minigrants are small grants designed to support the identification of commercially viable ideas and scientific work as well as encourage entrepreneurship among the scientific and SME community. The objective of a minigrant is to provide an initial financing grant to support entrepreneurs and SMEs in transforming basic ideas for innovative commercial activities into a business plan format that can be presented for consideration under a matching grants program and, if and when the project matures, to potential investors (angel or venture capital).

The purpose of a minigrant program is twofold: to stimulate the initiation of entrepreneurial activities in the field of innovation by providing small grants to help potential entrepreneurs take their existing ideas and determine whether these ideas can evolve into commercially viable ventures; and to help scientists and entrepreneurs who have limited experience in building successful companies obtain technical assistance and consultancy services that can help them conceptualize the business functions that would be needed to take their products to market.

Because many of the most innovative ideas evolve through the process of scientific research and because the focus and objectives of scientific research differ significantly from the processes involved in establishing and running a business, there is often a disconnect between accomplishments that are achieved in the laboratory and successful innovations emerging in the product and consumer marketplace. This disconnect is compounded in the ECA region, in which most countries have emerged just during the past 20 years from a centrally planned economic system and, as such, do not have a long-standing tradition of entrepreneurship. Consequently, the minigrants tool is designed to help stimulate the evolution of an entrepreneurial mindset and provide an incentive for scientists to innovate by offering them access to a vehicle that can help them gain commercial success following their success in the lab.

Minigrants also maximize the likelihood of success of given innovators by matching them up with additional skills and resources to which they likely would not otherwise have had access. Most innovators expend whatever limited resources they have in developing and building the technological value of their ideas—that is, working toward proving that their ideas are technologically feasible. This approach results in a knowledge gap, whereby insufficient time is afforded to assessing and documenting the commercial feasibility of their ideas. Furthermore, given that many of these individuals are likely to be inexperienced in taking on the tasks needed to help assess commercial viability (such as market analysis, marketing plans, and financial plans), access to this expertise helps improve the quality of the viability analysis.

A key drawback to minigrants, however, is the high degree of business support services required in the form of advice, knowledge, and technology transfer in the selection process and the grant implementation. Administering such a program, especially when the level of the capacity of the bureaucracy is low, will be challenging. In many ECA countries, the level of private business support services—such as consultants, training, business mentors, entrepreneurial networks, and even infrastructure services—is low. But public provision of business support services has generally proved ineffective. Thus, to be effective the financial instrument of minigrants must be combined with some way of providing business support.

A possible solution would be to design a combined system of minigrants with "virtual incubators." Recipients would receive the money only in conjunction with subscription to services of a virtual incubator. That would ensure that they receive the appropriate advice and mentoring, plus it would reduce the moral hazard problem of such a system. Virtual incubators are incubators without a physical infrastructure, which prevents the incubator from developing into a real estate business, a result that occurs only too often. It also would put more emphasis on providing business support services. That said, entrepreneurs will have to provide their own location and infrastructure might be a problem in some ECA countries.

Matching grants

Although the more traditional approach to R&D support to firms has been through tax incentives or subsidized loans, since the 1980s there has been an increasing awareness among OECD countries of the benefits of matching grants in encouraging firms to share and manage risk. A number of historically successful grant programs, such as Australia's R&D Start Program and the U.S. SBIR program, have an implicit matching component in that firms are expected to support a portion of the research budget. In countries such as Finland and Israel (box 4.3), more formal matching grant programs are helping to create a seedbed of precommercialization activities out of which the most promising innovations can be generated for follow-on investment by private sector investors, such as venture capital firms.

BOX 4.3
Do matching grants for industrial R&D help the Israeli economy?

In 2008, a study of the effects of government support to industrial R&D on the Israeli economy was undertaken at the request of Israel's Ministry of Finance and the Office of Chief Scientist at the Ministry of Industry, Trade and Labor. It uses econometric models to provide novel estimates of growth of industrial output as a result of government support to research and development (R&D), drawing on data from industry and R&D surveys in 1996–2003. The focus is on support to industrial R&D through direct grants to enterprises.

The key research questions are:
- Is government support replacing—that is, crowding out—private funding of R&D or is the support incentivizing additional private funding beyond what would have been undertaken without support?
- Is the government increasing output growth in the economy by increasing R&D? The study presents a methodology to take account of the dependence of the social rate of return on private returns as well as on spillover effects from the stock of R&D in one firm on the productivity of other firms in the economy.

The methodology consists of the following stages. First, propensity scoring to estimate the probability of receiving support in a probit model, using the model to create pairs of companies where one received a grant (treated) and one did not (control). Second, running a regression to analyze the difference between R&D expenditures in a company that received a grant compared with a control group company.

The crux of the matter for evaluating the effect of the R&D subsidy is to know what the firm would have spent on R&D had it not received the subsidy. As this counterfactual information is not available, the estimation methods used in this study essentially attempt to estimate the missing expected counterfactual by estimating the subsidy effect.

The key findings are:

- *Government support creates new R&D in Israel that would not have been undertaken had it not been for the support.* The estimates show that the scale of the new R&D is two to three times the amount of the (marginal) government support and that this effect is stable and significant across sectors both in industry and in the software and R&D sectors. The effect is similar for firms of different technology levels and firm size. Thus, it can be concluded that the current support instruments do not crowd out private funding.

- *A government grant leads to a more than double R&D increase.* In the industrial sectors, a government grant of NIS 1 million (around $280,000) creates a NIS 1.28 million increase of private R&D expenditure—which means a total increase of NIS 2.28 million in the magnitude of the economy's R&D. In the software and R&D sectors, the addition to private R&D expenditure is NIS 1.8 million, for a total NIS 2.8 million increase on R&D in the economy. These estimates are lower bounds to the actual additionality because they have been estimated for gross grants, without deducting repayment of royalties.

- *Government R&D support boosts industrial output.* The total effect of successful R&D consists of both a direct effect of the firm that conducted the R&D and an indirect effect on other firms (spillover). The study shows that the total return to the economy from government R&D support is very high—with spillovers coming mainly from medium to large firms (sales of NIS 50–300 million) and very large firms (more than NIS 300 million). The results indicate a multiplier of at least 5 to 6 between government support and the future total increment of the industrial output for medium to large firms and a multiplier of 1.5 to 2 between government support and the increase of industrial output.

Source: Lach 2002.

A matching grants program works by encouraging risk-sharing with firms, and it orients the selection process toward R&D programs that are most likely to generate innovations that can be commercialized. Qualifying firms, or consortia from academic institutions, submit grant applications for specific R&D projects that are reviewed by an independent research committee. If approved, the applicants receive a grant from 50 percent to 70 percent of the stated R&D budget for the project. Successful projects—that is, those leading to sales—will be required to repay the grant, as a royalty from revenue, up to the dollar-linked amount of the grant. The sharing of risk with the firm alleviates, though it does not eliminate, the negative consequences of "picking winners" by the public sector. The royalty scheme also orients the selection process toward picking projects mainly to achieve sales and profitability targets.

There are two critical aspects of a matching grant program that make it useful in the ECA context. First, firms are required to invest a dollar of their own funds for every dollar they receive as a grant. Proof of the private expenditure of a dollar is required before the government reimburses the entrepreneur for the dollar it invested. The importance of matching stems from the fact that its effect is to reduce the marginal cost of research to the firm, which will provide an incentive to increase total expenditure, precluding the dollar-for-dollar crowding-out result.

Second, the administration of matching grants must involve an independent and effective selection process whereby the projects most likely to generate commercial innovations are chosen. That factor is crucial. In the ECA region the selection mechanism will face the risk of excessive administrative burden and corruption. As much as possible, industry experts and private sector players who are familiar with commercialization of innovations should be involved in the selection. It is also important that the criteria for selecting projects and using grant proceeds are clearly laid out and adequate to the country environment.

Matching grants (as well as minigrants) also have the potential to create and foster links between the private sector and universities and research institutes by favoring consortia. Cooperative schemes between the private and public sector have been at the heart of many programs in OECD countries. Because the main aim of these instruments will be to promote commercial innovation, it is also important that the private sector be in the driver's seat of these consortia.

Venture capital support

Although matching grants support ESTD, venture capital plays a vital role in the commercialization phase. It targets high-risk and high-return projects that have passed the early stage whether they have been supported by a grants program. To achieve the high commercial returns expected by investors, venture capital seeks out companies that have successfully developed the innovation, proved its technical capability, and identified probable commercial applications and markets. At that stage, venture capital provides the funds to expand production, develop markets and the customer base, and play a critical role in supporting the later (and most visible) stages of commercialization. As chapter 1 mentioned, new enterprises, particularly those backed by venture capital, have proven to be a key engine for innovation. Whereas large firms often focus on existing clients and markets, new companies will often focus on exploiting new market opportunities.

But venture capital also has its limitations. First, although it plays a role in financing the commercialization of innovation and the expansion of innovative firms[46]—and the success of venture capital funds depends on having a "deal flow" of attractive companies—it does not provide a solution to the market failure in ESTD. Thus, any intervention supporting venture capital needs to be preceded or complemented by interventions addressing the early funding gap through matching grants or by other means.

Similarly, minigrants and matching grants programs are likely to work best in circumstances in which support for later stages of the innovation process is available. Entrepreneurs will be able to plan better, and their incentives for engaging in commercial R&D will be greater if they know that there are adequate support instruments available after the initial stages of the innovation process. Venture capital measures should also be coupled with reforms to improve the conditions for developing a venture capital industry, including further revising venture capital legislation and increasing the stock market's liquidity. The countries with some of the most robust venture capital industries (Australia, Canada, Israel, and the United States) have active programs at all stages of the innovation life cycle: from grants to venture capital support programs. Australia has the R&D Start Program to provide grants for the commercialization of innovations by SMEs, which is complemented by its Innovation Funds Program to encourage venture capital investment in innovative SMEs.

A second limitation is that venture capital cannot be relied on to provide all of the financing necessary for innovation. Only 1 in 200 SMEs in emerging markets (compared with 1 in 100 in the United States) is able to secure venture capital financing (Nastas 2005). In fact, in middle-income countries, large multinational corporate investors (for example, Shell, General Electric, IBM), rather than venture capital investors, are often best suited to provide access to finance for innovation. Multinational corporations can coinvest to form public–private funds locally, particularly to finance projects in sectors of the economy in which these corporations are interested in developing new startups to strengthen the supply chain. The latter may be of particular relevance for countries rich in natural resources, such as Kazakhstan, the Russian Federation, and Ukraine. Corporate ventures will become a growing source of ESTD financing if and when local corporations accumulate sufficient retained earnings to allow them to engage in risky ventures that may be only tan-

---

46. Venture capitalists act as the first step of formal commercial financing for innovative, high-growth firms. They often play a similar role as angel investors in providing critical technical assistance and managerial support. Venture capitalists also support firms in accessing later rounds of financing, including initial venture capital public offerings.

gential to their core business. Ideally, such new ventures will be spun off to establish more flexible and entrepreneurial SME startups—assuming there are farsighted entrepreneurial owners in these private companies.

What lessons can we glean from the OECD countries? The success of the most prominent venture capital funds relies on three characteristics:

- *Investment expertise.* Venture capital investment analysts are highly specialized, with a strong understanding of different technology fields and their markets. If a venture capital fund invests in a company, it typically gains high levels of control and influence over the company's management decisions. The venture capital fund manager brings management and commercialization expertise and exercises control to ensure commercial success.

- *Risk profile.* Venture capital investment strategies are formulated such that they can absorb a high number of failed investments. Typically, the venture capital fund aims to earn very high returns from 1 or 2 of 10 investments it makes, which compensates for the expected failure of the rest of the investments (cross-subsidization).

- *Deal flow.* Venture capitalists rely on a supply of high-potential companies emerging from the earlier stages of business, technological, or innovation development. Thus, venture capital works best in economies (such as Israel and the United States) in which the early stage of technological development is financed by internal funding, angel investors, or government-supported grant financing.

These three characteristics, however, constrain the possibility of government intervention. And misperceptions of the role of venture capital have led to a number of failed interventions in the risk capital markets.[47] State-owned and state-managed venture capital, in particular, have proved to be especially prone to failure. Government officials usually do not have the crucial technical expertise and risk-taking mindset to support innovations at their commercialization stage. Thus, caution is advisable in the latest Russian government's initiative to establish "private–public early stage regional venture capital funds." As in fact is planned, the private sector's participation is critical for success.

Typically, in many cases in which so-called venture capital funds managed by government entities are operating with commercial success, the risk profile does not display features of venture capital and the funds are not being invested in innovative ventures, but rather in small, more

---

47. Because efforts to promote the emergence of a venture capital industry have failed in at least one reform-oriented country (Chile), a careful analysis of the reasons for this failure and any design problems that may have caused the failure is needed.

mature companies with less risky product lines. Capture and rent-seeking are prevalent and problematic because these types of fund setups are prone to be dominated by political interests (patronage).

Even so, there are a number of successful examples—in Israel, the Republic of Korea, Taiwan, China, and the United States—in which government support for the development of a private venture capital industry has played a significant role in the development of a dynamic innovative sector. In these cases, the government has "seeded" the venture capital industry through investing in privately managed funds. Governments mitigate some of the risk inherent in technology-oriented SME startups, and the venture capitalist provides commercial and managerial expertise. In time, funds graduate from government support to avoid restrictions placed on the fund. This type of instrument can take many forms:

Direct co-financing. By participating in a privately managed venture capital fund, the government lends credibility to the fund and acts as a catalyst for other investors to participate. This works well if the venture capital industry is experienced and there are attractive opportunities. Take the case of Israel's Yozma Fund. In 1992, the government provided $100 million divided among 10 private funds. Each fund manager raised a matching amount of private funding. The funds made investments of $300,000–750,000 in hundreds of companies. By 1997, the government felt that it had achieved its goals and sold the Yozma Fund through privatization.

Leveraged returns. In this scheme the government, either by subscribing for ordinary equity shares or providing grants, coinvests with private investors but takes only a small part of the return—thus "leveraging" the upside potential for private investors. The Australian Innovation Investment Fund Program provides up to two-thirds of the capital for the venture capital funds but takes only 10 percent of the return with the remaining 90 percent allocated to private investors and management. In exchange, fund managers are required to invest a portion of their fund in SMEs and early stage companies. Israel's Yozma Program and the U.S. Small Business Investment Company program have variations on this basic approach. These programs have proved successful in countries in which there are opportunities to achieve high returns.

Guarantees. Guarantees against losses have been successfully used to promote investments in venture capital funds but tend to be most useful for countries with financial systems capable of sophisticated risk evaluation. By guaranteeing a certain return to investors or taking a subordinated position in the distribution of the funds' profits, the government protects investors against major losses of principal (downside risks are capped). Although guarantee programs can mitigate risk and attract commercial capital, they sometimes distort investment decisions. Facing limited losses, venture capitalists tend to be less rigorous in assessing the downside of deals.

In the ECA region—where venture capital is still quite limited— venture capital leveraging may be the most effective approach. The government may need to provide incentives for private investors beyond merely cofinancing venture capital funds. However, guarantee programs are more complex to implement, and ECA governments are unlikely to have the capacity to effectively evaluate the guarantee risk associated with the funds.

Institutional support instruments

Even if researchers or inventors have the capacity to innovate, they may lack the skills, knowledge, and business acumen to develop a project that would be acceptable for funding, not to mention the ability to engage in the business planning and implementation necessary to commercialize their innovations. In that case, the deal flow and utilization of the instruments may be constrained by the lack of project or business development capacity.

Many OECD and developing country governments implement business support services, especially incubators, as an instrument to promote the commercialization of innovation (box 4.4). The rationale is that innovators and nascent entrepreneurs cannot be expected to manage all parts of the commercialization process—from launch strategy to financial planning, to intellectual property rights, to basic logistics. By providing logistical support and technical assistance, incubators are meant to help entrepreneurs transition from a supportive institutional environment of universities or large companies to the more challenging environment of a new company or R&D project.

BOX 4.4
Armenia's efforts at enterprise incubation

Armenia's prospects for inclusive and long-term growth require it to diversify into knowledge-driven sectors and upgrade the quality of domestic products and services. As a land-locked country Armenia is subject to transport cost disadvantages that can only be compensated by moving to higher value products and services. At the same time Armenia is plagued by the absence of support structures, lack of value chains, and low appreciation for the value of intangibles (brand names, business reputation, marketing and managerial skills, networks, and so on), which hamper all high-tech industries, including information technology (IT).

In 2000 the government of Armenia declared the information and communication technology (ICT) sector a priority. It also approved an ICT Master Strategy to help ICT firms become more competitive in overseas markets. To implement the strategy, Armenia, in partnership with the World Bank, set up the Enterprise Incubator Foundation as an IT business development and incubation agency. The nonprofit agency focuses on three key areas:

- Business linkages services. The Business Services Center provides specialized business development services to interested clients, mostly locally owned software companies without the necessary management and marketing skills. It also places managers and senior specialists of Armenian-owned software companies into IT companies and incubators in more advanced market economies.
- Skill development services. The Skills Development Facility—the largest project component given substantial skills shortages—provides advanced models of continuing IT professional education through local and international training, internships, seminars, and competitions.
- Managed workspace. The Facility Services provides communications infrastructure and office space for lease to IT companies—domestic and foreign, already operating or startups.

How has the Foundation performed? One of its key achievements is highlighting that an advanced incubation model could accelerate the development of the IT sector, with economic spillovers for other manufacturing sectors. Besides tackling the sector's infrastructure needs, it provided business development and training services to about 109 companies, or around 73 percent of the estimated 150 companies active in Armenia's IT sector. The main beneficiaries were startup and nascent companies—about 69 percent of the firms had fewer than 50 employees and 43 percent had fewer than 20 employees—though the Foundation also provided services to institutes, universities, and a large telecommunication company. The Foundation reports that by 2010 there had been about a $1 million increase in sales contracts for the local IT companies annually and a 10 percent growth in IT exports annually.

The Foundation has designed and implemented several large public-private partnerships. These include: setting up a Microsoft Innovation Center in Armenia jointly with Microsoft and the U.S. Agency for International Development, a "Computer for All" project jointly with Hewlett Packard and Unicomp CJSC, and an advanced educational and technology resources program with Sun Microsystems. In addition, the Cafesjian Family Foundation established a venture capital firm with a statutory fund of about $1 million.

Going forward, Armenia, which was hard hit by the global financial crisis, hopes to position itself to take advantage of the global economic recovery. High on its policy list is strengthening competitiveness and growth—which means continuing to develop its ICT sector to facilitate an information society and knowledge-based economy.

Sources: Economy and Values Research Center 2008; World Bank 2007, 2010a; World Economic Forum 2010; updated statistics from the Enterprise Incubator Foundation.

However, incubator programs have often received poor marks for effectiveness, both in terms of their success in promoting businesses and cost-effectiveness. This is in part because many incubator programs have devolved from the original goals of business support to the provision of real estate and office support services. Although there has been relatively little analysis or impact assessment of these types of programs, the reasons for the perceived lack of success include the following:

- *Necessity for specialized skills and knowledge.* The technical advice or market knowledge required by inventors and entrepreneurs is often highly specialized, whereas most technical assistance programs are designed to reach a range of business needs and therefore are general in nature. The type of business advice needed, however, is from seasoned, experienced business people who are unlikely to be consultants to these types of programs. Therefore, the skills and knowledge required by an entrepreneur in a specific context rarely match the technical assistance provided by a program.

- *Government allocation of resources.* These types of programs tend to be highly subsidized by government and designed and managed by government agencies or public universities (supply driven)—without necessarily taking inventors' needs into account. Public officials are ill-suited to evaluating the paradigm of inventors or entrepreneurs and effectively allocating resources to support them.

- *Incentives for entrepreneurs.* The entrepreneurs who are most likely to be successful may not have a need for "incubation" because they have the wherewithal to mobilize needed resources and directly confront the seasoning process of becoming an entrepreneur, which requires overcoming the challenges of commercializing an innovation and setting up a company. Thus, there is a self-selection process whereby weaker candidates access incubation services and are unable to graduate from those services.

- *Financial sustainability.* Governments often expect incubators to quickly become financially sustainable. This leads to no or little operational budget for the "soft" aspect of incubation, once the budget has been spent on its "hard" infrastructure. Incubators must then rely on attracting mature, low-risk companies for a steady flow of income and cutting back on service provision.

The result is that many existing business incubators and technology parks in post-transition countries, including Russia, are little more than custodial care centers and, in the worst case, tax havens. A recent study argued that "elements of infrastructure such as technological parks and innovation and technology centers are considered by managers of small companies more as nice premises, with subsidized rents, rather than structures enabling to promote small enterprises renting these premises" (Gaidar, Sinelnikov-Mourylev, and Glavatskaya 2005). They are primarily controlled work premises designed to help startup firms survive in the midst of a hostile environment. There is a logic to some form of infrastructure support in hostile environments in which land is difficult to rent, utility and communication connections are difficult to organize, and petty harassment (or worse) from bureaucratic inspectors is an unfortunate but common fact of life. But these real estate services should not be confused with the proactive value added support and the tools, information, education, contacts, advice, and resources critical to success in the ESTD of a business.

Given the problems identified in government-led provision of support services, it is necessary to critically evaluate direct subsidies for these types of programs and to be more creative in developing solutions to the real challenge of delivering effective managerial and technical business support to ESTD companies.

It can be argued that the most successful models for incubation and business support services for new companies are the angel investors and early-stage venture capital funds that operate primarily in the U.S. market. Typically, these angel investors are seasoned business people with experience in the industry, who provide business advice and contacts for specialized skills and knowledge. Because they are investors in the business, they resolve the failures noted above by having the incentive to do rigorous due diligence on the capabilities of the entrepreneur, to critically evaluate the needs of the business and direct it toward necessary resources, and to remain involved in the company, "hand holding," over an extended period of time.

Unfortunately, angel investors are not prevalent in the ECA. Even so, a number of best practices can be taken from the model and applied to providing business support services:

- *Business support services linked with investment activities.* Business support services should be a paired component of an investment (such as a matching grant or venture capital investment). The entrepreneur is thus screened as part of the investment selection process, and the investment capital helps the company execute its business plan.

- *Private allocation of support services.* Rather than having the government set up ponderous technical assistance schemes, entrepreneurs articulate and determine the type of assistance they need as part of their application for a grant or venture capital. A portion of the investment budget is allocated specifically for business support services.

- *Private provision of technical skills and knowledge.* Entrepreneurs are empowered to seek out the skills and knowledge they need from private consultants—making the supply of business services demand-driven.

- *Seasoned business expertise.* As much as possible, successful business people in a country or region should be involved in the design and selection process of innovation programs. Even though they are unlikely to become intimately involved in the selected businesses, their advice and perspectives can be diffused throughout the program.

In recent years, OECD countries have experienced a burst in mentorship-based programs and entrepreneurship support programs. These have ranged from public programs sponsored by governments or universities (for example, the MIT Venture Mentoring Service in the United States) to private programs revolving around business associations (for example, The Indus Entrepreneurs, globally) to for-profit programs backed by angel investors or venture capitalists that take the form of "startup accelerators" (for example, TechStars in the United States). Mentorship programs typically provide startup teams with access to a community of serial entrepreneurs and investors who coach them on a variety of business issues, provide them with feedback on their business ideas, and help them network with relevant market players. Because mentorship programs focus on soft support, much of which is pro bono, operational costs are low and organizational structures are lean allowing programs to experiment with a range of approaches.

The bottom line is that business support services should be viewed as a complementary support to the core instruments that provide financing for innovation and R&D, not as a standalone policy lever. Business support services should be supplied in a demand-driven way, with financial assistance for business support services being offered to firms—who in turn will use the financial support to purchase business support services.

Monitoring and evaluation

An effective monitoring and evaluation (M&E) framework—one that is flexible enough to take into account the organic nature and unpredictability of innovation policies—should be a component of the deployment of policy instruments targeted at enhancing innovation. There is likely to be a significant amount of uncertainty about key variables (for example, the pipeline of the project). There may be time lags because the policies change the incentives related to innovation activity. The dynamic nature of the political and business environment will also affect the uptake and performance of the program.

The role of the M&E system is to identify how a program is performing relative to its objectives to allow correction or cancellation of the program midcourse, for example, as a result of low uptake (the project proposal pipeline is lower than expected). Specifically, an M&E system should follow up on whether the program is overfunded and unnecessarily tying up resources and evaluate the overall success of the program on an ongoing basis to help determine when and if the program should be discontinued or restructured. To fulfill that role, M&E vehicles must effectively provide data about the program's progress on three levels—project, program, and economy.

Project. The M&E framework should provide feedback on whether the tools being deployed are making a significant impact on the projects being financed. That is, are they increasing the projects' likelihood of success and their access to follow-on financing? To accomplish this, the projects must be evaluated on two dimensions: technological success and commercial success. This process begins with ensuring that projects submitted for evaluation for funding clearly state, and distinguish between, the technological and commercial objectives. These objectives should be approved by the evaluation committee or authority and should be amended in conjunction with the applicant where necessary. In doing so, each project begins with specific, measurable, time-bound benchmarks that can be evaluated to assess the overall success of the project.

Although measuring how a project performs relative to its benchmarks makes it possible to determine whether the project has been a success, it does not effectively convey whether the program itself has had a significant contribution to the success or whether the program merely identified which projects were most likely to succeed. In other words, were these projects successful because the tools made available facilitated this success or because the selection process identified the best projects, with these projects likely to produce a successful outcome independent of the support provided by the program itself?

Jaffe's "regression discontinuity design" (2002) can be an effective tool in helping to answer that question, as it focuses on evaluating projects at two stages: the selection stage and the assessment stage. The tool requires that projects be ranked in relation to their likelihood of success at the selection stage (a process that needs to take place in any event to enable the selection of projects for funding) and at the assessment stage, to include not only those projects that were selected for funding but also those not selected for funding. Essentially, the evaluation team will keep track of projects that were not selected for funding and apply this tool to assess to what extent the provision of funding and innovation support contributed to the increased success of the selected projects.

Program. The M&E framework also needs to answer whether the program succeeds in directly stimulating private investment in innovation. Many R&D grants programs, technical assistance programs, and venture capital programs have suffered from poor utilization of funding because the program was too early in the life cycle of innovation, or because of poor program design or poor implementation. An M&E tool, therefore, should aim to monitor the use of funding and the underlying factors driving use. Low uptake during the early years of the program should be expected and even encouraged. Above all, pressure to invest in lower quality projects during this early phase should be avoided because early failures are much more likely to hurt market perception of investments in innovation than is slow ramp up.

At the same time, underuse of funding must at some point be evaluated to determine whether to release funding for other uses. In that regard, the program design in relation to the attractiveness of the financing package to the private sector, as well as the implementation vehicle, should be evaluated to determine whether they are creating bottlenecks in the program. Surveys of program firms/clients and nonclients can provide input on whether and how the program is fulfilling its objectives. For example, the selection and disbursement process should also be evaluated in regard to its efficiency, effectiveness, and burden on the applicant. Overuse of funding can be evaluated similarly as to whether the incentive structure is so attractive that it leads to crowding out of private investment and whether the selection process is consistent with the objectives.

Economy. The most difficult determination will be whether the overall program was beneficial for the economy. Did it help to generate greater returns in relation to increasing the productivity and growth prospects of the local economy? M&E at the economy level is complex because the evaluation must show not only whether productivity and growth have increased but also whether the program had an impact on that increase. Because publicly funded R&D should be targeted at the generation of spillovers, it is important to focus on both the commercial or technological successes under the program and whether the program helped to generate increased R&D beyond the sum of its projects and to contribute to innovation at the economy level. Then, too, identifying the causal connection between the specific program and the changes in the economy requires a long time span: ex post evaluation will have to take place when a sufficient number of years have elapsed since the beginning of the program.

An attempt to identify the causal connection between the specific program and economic changes has been conducted by the Israeli government. The results shows that the total return to the economy from government support to R&D is very high—with increases in industrial output, no crowding out, and a government grant leading to a more than doubling of R&D expenditures (see Israeli case study).

Conclusion

There are numerous challenges ahead for ECA to ignite its latent scientific and technological potential and regain the lead in regional or even global settings. The countries in the region differ in their remaining reform agendas, as well as in the relevant resource endowments needed to transition toward a more dynamic and globally competitive knowledge economy. The rapid technological catchup in the more reformed economies indicates that within-firm productivity improvements, rather than reallocation of resources, are increasing in importance as a source of growth in ECA. What is the role of government to boost knowledge absorption in ECA? Our book suggests focusing on the following objectives:

- Creating a favorable investment climate, as well good institutions—such as national education and R&D systems—and government support instruments that help firms and researchers overcome market failures that dampen the culture of innovation and entrepreneurship. After all, the process of innovation is neither automatic nor costless.

- Opening more to FDI to encourage knowledge absorption. Countries that are poor in natural resources, will have a hard time attracting greenfield FDI unless they improve their investment climate—especially in areas such as starting a business, protecting investors, getting credit and enforcing contracts.

- Intensifying international R&D collaboration and foreign R&D investment to enhance ECA's integration into the global R&D community. During this process, governments should support local coinventors in obtaining international patent protection before they negotiate the ownership of their joint patents with their western coinventors.

- Restructuring RDIs in a manner that takes into account the real potential of each organization and the local and global demand for its outputs.

- Evaluating and redesigning financial support programs—and developing new institutions and instruments based on international good practices (matching grants, minigrants, venture capital, parks, and incubators)—to tackle key pressure points along the innovation and commercialization continuum.

The good news is that igniting innovation and technology adoption so this becomes a central part of the ECA countries' development and growth strategies is within reach.

Case study: How Israel has promoted innovation in recent decades

In 1968, the Kachalsky Committee appointed by Prime Minister Levi Eshkol recommended that Israel increase dramatically the amount of R&D activity performed by the private sector. The Office of the Chief Scientist (OCS), established in 1976 to implement the recommendations, received a mandate to subsidize commercial R&D projects undertaken by private firms. Until then, support had been confined to R&D Labs and academic R&D, in addition to the weighty resources that were devoted to defense-related R&D and agricultural research.

As a recent rigorous impact study shows, the stepped up OCS support paid off. Between 1969 and 1987, industrial R&D expenditures grew at 14 percent a year, and high-tech exports grew from a mere $422 million in 1969 (in 1987 dollars) to more than $3 billion in 1987 (Toren 1990) and to $14 billion today—about half the industrial exports of goods—and they proved very resilient during the ongoing global downturn. Multinational companies have also established R&D centers—initial entrants included Motorola (1964), IBM (1972), Intel (1974), and National Semiconductors (1978), and more recently Google, SAP, and Oracle, among others. But government support for R&D can do little in the absence of human resources and a friendly investment climate. On this score, Israel was helped by the massive immigration from the former Soviet Union of about 1 million immigrants, many of whom had advanced technological educations. Two key factors—ranked highly in the World Bank's Doing Business project—were instrumental: deregulation and liberalization of Israel's financial markets; and openness to trade and investment (including the worldwide information and communication technology boom and the globalization of U.S. capital markets with respect to financing startup activity and initial public offerings).

The Matching Grants Program

At the heart of Israel's R&D support schemes stands the OCS with an overall budget of $300 million a year. Its main activity is industrial R&D fund, providing matching grants of typically 50 percent to selected industrially competitive projects. This program has no specific priorities and selects projects according to their merit. All firms intending to export part of the project's outcome qualify. Thus, upon approval of the project, the firm commits to match, dollar-by-dollar the subsidy received from the OCS. And companies are obliged to pay royalties to the OCS when the project succeeds (that is, leads to sales). They are also prohibited, under the Israeli R&D law, to sell their technology abroad unless the OCS is fully compensated. Qualifying firms submit grant applications for specific R&D projects, which are reviewed by a Research Committee, and if approved (about 70 percent are), the applicants receive a grant of up to 50 percent of the stated R&D project budget. Successful projects are required to repay 3–6 percent of their sales in royalties, up to the dollar-linked amount of the grant.

In 1993, the OCS established the Magnet program, with a mandate to support the creation of consortia made up of industrial bodies and academic institutes to develop precompetitive, generic technologies. The rationale was that public support was needed to narrow the gaps between Israel's world renowned academic institutes and local industry. It was also a reaction to the growing phenomenon of these programs in Japan, the United States, and EU countries.

The Yozma Program

Toward the end of the 1980s, the OCS realized that the conditions for the creation of a venture capital industry were in place—such as technological capabilities that were developed as a result of substantial investments in applied research by both the public and private sectors (for example, OCS matching grants projects).[48] For that reason, Yozma was organized in the early 1990s as an independent entity under contract to the OCS. The plan was that the state would withdraw from the program after seven years and the investors' team must include a foreign partner with expertise in venture capital investments and a local financial partner.

What lessons can we draw from Yozma's two decades of experience when considering the creation of a venture capital industry in another region?[49]

---

48. The discussion of venture capital is based on Teubal and Avnimelech 2002.
49. Modena 2002.

- *It is best to involve foreign partners.* The Yozma Program's requirement to involve experienced foreign partners in the funds led to the recruitment of some of the most important venture capital investors worldwide. The local emerging high-tech industry benefited greatly from the image they provided and their vast experience and extensive international networks.

- *Public intervention is a useful trigger to create a venture capital industry.* Most startup capital is derived from international funds, unlike seed capital, which requires continuous governmental provision. The Yozma Program was initiated in 1993 and by 1998 all the funds had been privatized. Private investors felt more confident knowing that the government's involvement would be limited to a short period of time and that a specific exit date was already determined.

- *The state should be a passive investor.* The state representative on the boards of the funds refrained from interfering in the investment decisions to allow market-oriented decisions and only assured that the fund was acting according to regulations. This decision freed the funds from unnecessary bureaucracy and allowed them to operate according to the markets needs.

- *Upside incentives have their virtues.* Upside incentives (incentives that motivate funds to be more profitable in case of success) appear to be more successful than downside incentives (incentives that limit the investors' losses in case of bad investments.

- *Avoid giving a monopoly to any one fund.* It is important to make sure that public funds are controlled by different management companies to avoid monopolization of the funds.

Technology Incubators Program

The Technological Incubators Program (TIP) was established by the OCS in 1990 following the mass immigration of highly qualified personnel from the former Soviet Union (though it was also open to entrepreneurs born in Israel). The program aimed at assisting the immigrant scientists and engineers to develop ideas while being (self) employed, in the short term, and potentially creating new high-tech companies and more jobs in the mid to long term.

Within the project's first two years, 28 incubators were established in peripheral regions and cities. The incubators were founded in cooperation with universities, local authorities, and large firms. Each was founded as a not-for-profit entity providing financial support, consulting, and office and lab space to about eight incubator companies each year.

In 2002, the OCS began privatizing incubators, and by the end of 2006, 17 of the existing 24 incubators had been privatized, and 1 new private Bio-Tech incubator was established. About 200 projects in various stages of R&D are being carried out in the TIP at any given time.

What lessons can be drawn from TIP's experience?[50]

- *Strong public support for seed finance is vital.* Only 2.4 percent of the incubator projects received funding from venture capital while still in incubation, a very low proportion considering that Israel's high-tech industry relies heavily on venture capital funding—52 percent of firms are funded by venture capitals.

- *Strong initial public support can decrease as projects mature.* Private funding to the incubator management team increased over time, and over the years most of the incubators were privatized. Today 17 of the 24 are privately managed (though government money is still invested in all the incubators). Strong public support may be needed at the beginning but can be reduced as the program matures and shows success.

- *Expert networks should perform evaluations.* The TIP has succeeded in establishing a network of experts who assist the program in their selection process, which is vital given that projects submitted are from varied fields and reflect very specific expertise.

- *The entrepreneur's share plays a big role.* The TIP stipulates that at least 30 percent of the shares of the company (after the first round of funding) remain the property of the entrepreneur. This keeps the entrepreneur motivated and strongly involved in the company the entrepreneur has the most technological know-how needed for developing the company.

- *Great care should go into choosing a manager.* The motivation and capability needed to assist the growing enterprises are just as important as those needed to run an existing enterprise. Thus, strong attention should be given to choosing the incubators' leading personnel. The privatization process dramatically contributed to raising the competence of the incubators' management.

- *Incubators need to be close to universities.* Shefer and Frenkel (2002) have shown that the proximity of an incubator to a university research center is of great importance especially in the life science fields.

---

50. Modena 2002.

Summary

The preconditions in place in the early 1990s that fostered the development of the Israeli high-tech industry included:

- The availability of experienced scientists and engineers seeking work (either immigrants or persons having been discharged from the defense industry), with a strong motivation to initiate new enterprises.

- A general positive overview of entrepreneurship by the society; in many cases entrepreneurs were perceived as role models.

- Deregulation and a liberalization of Israel's financial markets, which spawned a better environment for doing business.

- The beginning of the globalization of the U.S. capital markets with respect to financing startup activity and initial public offerings.

References

Acemoglu, Daron, and Fabrizio Zilibotti. 2001. "Productivity Differences." *Quarterly Journal of Economics,* 116 (2): 563–606.

Acharya, Ram C., and Wolfgang Keller. 2008. "Estimating the Productivity Selection and Technology Spillover Effects of Imports." NBER Working Papers 14079. National Bureau of Economic Research, Cambridge, MA.

Aghion, Philippe, and Peter Howitt. 2005. "Appropriate Growth Policy: A Unifying Framework." 2005 Joseph Schumpeter Lecture delivered to the 20th Annual Congress of the European Economic Association, Amsterdam, Netherlands, August 25.

Aghion, Philippe, Nick Bloom, Richard Blundell, Rachel Griffith, and PeterHowitt. 2005. "Competition and Innovation: An Inverted U Relationship." *Quarterly Journal of Economics* 120 (2): 701–28.

Aghion, Philippe, Wendy Carlin, and Mark Schaffer. 2002. "Competition, Innovation and Growth in Transition: Exploring the Interactions between Policies." William Davidson Working Paper 501. William Davidson Institute, University of Michigan Business School, Ann Arbor, MI.

Alesina, Alberto, Silvia Ardagna, Giuseppe Nicoletti, and Fabio Schiantarelli. 2005. "Regulation and Investment." *Journal of the European Economic Association* 3 (4): 791–825.

Alam, Asad, Paloma Anós Casero, Faruk Khan, and Charles Udomsaph. 2008. *Unleashing Prosperity: Productivity Growth in Eastern Europe and the Former Soviet Union.* Washington, DC: World Bank.

Ali-Yrkkö, Jyrki. 2004. "Impact of Public R&D Financing on Private R&D— Does Financial Constraint Matter?" ETLA Discussion Papers 943. Research Institute of the Finnish Economy, Helsinki.

———. 2005. "Impact of Public R&D Financing on Employment." ETLA Discussion Papers 980. Research Institute of the Finnish Economy, Helsinki.

Arnold, Erik, Howard Rush, John Bessant and Mike Hobday. 1998. "Strategic Planning in Research and Technology Institutes." R&D Management, 28 (2): 89–100.

Arnold, Erik. 2007. Governing the Knowledge Infrastructure in an Innovation Systems World. http://idbdocs.iadb.org/wsdocs/getdocument.aspx?docnum=976011.

Arnold, Jens, Beata S. Javorcik, and Aaditya Mattoo. 2007. "Does Services Liberalization Benefit Manufacturing Firms? Evidence from the Czech Republic." Policy Research Working Paper 4109. World Bank, Washington, DC.

Athukorola, Prema-Chandra. 2006. "Trade Policy Reform and Structure of Protection in Vietnam." World Economy, 29 (2): 161–87.

Auerswald, Philip E., and Lewis M. Branscomb. 2003. "Valleys of Death and Darwinian Seas: Financing the Invention to Innovation Transition in the United States." Journal of Technology Transfer, 28 (3–4): 227–39.

Baumol, William J. 2002. The Free-Market Innovation Machine: Analyzing the Growth Miracle of Capitalism. Princeton, NJ: Princeton University Press.

Bernard, Andrew B. and J. Bradford Jensen. 1999. "Exceptional Exporter Performance: Cause, Effect, or Both?" Journal of International Economics, 47 (1): 1–25.

Blomstrom, Magnus, and Ari Kokko. 1999. "How Foreign Investment Affects Host Countries." Policy Research Working Paper 1745. World Bank, Washington, DC.

Bogetic, Zeljko, Karlis Smits, Nina Budina and Sweder van Wijnbergen. 2010. "Long-Term Fiscal Risks and Sustainability in an Oil-Rich Country." Policy Research Working Paper 5240. World Bank, Washington, DC.

Brahmbhatt, Milan, and Albert Hu. 2010. "Ideas and Innovation in East Asia." World Bank Research Observer, 25 (2): 177-207.

Brahmbhatt, Milan. 2007, Comment on Dani Rodrik's "Growth after the Crisis", Paper prepared for the Commission on Growth and Development. World Bank, Washington, DC.

Branstetter, Lee, and Nicholas Lardy. 2008. "China's Embrace of Globalization." In Loren Brandt and Thomas Rawski, (eds.), China's Economic Transition: Origins, Mechanisms, and Consequences, Cambridge University Press.

Branstetter, Lee G., and Mariko Sakakibara. 2002. "When Do Research Consortia Work Well and Why? Evidence from Japanese Panel Data." American Economic Review, 92 (1): 143-159.

Bresnahan, Timothy, and Manuel Trajtenberg. 1995. "General Purpose Technologies: Engines of Growth." Journal of Econometrics, 65 (1): 83-108.

Chandler, Alfred D., Jr. 1977. The Visible Hand: The Managerial Revolution in American Business. Cambridge, MA: Harvard Belknap.

Chen, Derek H.C., and Carl J. Dahlman. 2004. "Knowledge and Development—A Cross-Section Approach." Policy Research Working Paper 3366. World Bank, Washington, DC.

Chen, Derek H.C., Maggie Xiaoyang, Tsunehiro Otsuki, and John S. Wilson. 2006. "Do Standards Matter for Export Success." Policy and Research Working Paper 3809. World Bank, Washington, DC.

Clerides, Sofronis, Saul Lach, and James Tybout. 1998. "Is Learning-by-Exporting Important? MicroDynamic Evidence from Columbia, Mexico and Morroco." *Quarterly Journal of Economics,* 113 (3): 903–47.

Coe, David T., and Elhanan Helpman. 1995. "International R&D Spillovers." *European Economic Review,* 39 (5): 859-887.

Coe, David T., Elhanan Helpman, and Alexander W. Hoffmaister. 1997. "North–South R&D Spillovers." *Economic Journal,* 107 (440): 134–49.

Cohen, Wesley, and Daniel Levinthal. 1989. "Innovation and Learning: The Two Faces of R&D." *Economic Journal,* 99 (397): 569–96.

Commander, Simon, and Jan Svejnar. 2011. "Business Environment, Exports, Ownership, and Firm Performance." *Review of Economics and Statistics,* 93 (1): 309-337.

Corbett, Charles, María Montes-Sancho, and David Kirsch. 2005. "The Financial Impact of ISO 9000 Certification in the United States: An Empirical Analysis." *Management Science,* 51 (7): 1046–59.

Correa, Paulo, Liljana Tarade, and Iwona Borowik. 2010. "Croatia's Science and Technology Project Unleashes Innovation." ECA Knowledge Brief 24. World Bank, Washington, DC.

D'Souza, Juliet, and William L. Megginson. 1999. "The Financial and Operating Performance of Newly Privatized Firms during the 1990s." *Journal of Finance,* 54 (4): 1397–438.

de Ferranti, David, Guillermo E. Perry, Francisco Ferreira, Michael Walton, David Coady, Wendy Cunningham, Leonardo Gasparini, Joyce Jacobsen, Yasuhiko Matsuda, James Robinson, Kenneth Sokoloff, and Quentin Wodon. 2003. *Inequality in Latin America and the Caribbean: Breaking with History?* Washington, DC: World Bank.

De Loecker, Jan. 2010. "A Note on Detecting Learning by Exporting." NBER Working Papers 16548. Cambridge, MA: National Bureau of Economic Research.

Dainton, Frederick. 1971. "The Future of the Research Council System." In *A Framework for Government Research and Development.* London: Her Majesty's Stationery Office.

Dasgupta, Partha and Paul A. David. 1985. "Information Disclosure and the Economics of Science and Technology," CEPR Discussion Paper 73. Center for Economic and Policy Research, London.

David, Paul A. 2004. "Can 'Open Science' be Protected from the Evolving Scheme of IPR Protections?," *Journal of Institutional and Theoretical Economics,* 160 (1): 9-38.

Desai, Raj M., and Itzhak Goldberg. 2008. *Can Russia Compete: Enhancing Productivity and Innovation in a Globalizing World*. Washington, DC: Brookings Press.

Dutz, Mark, ed. 2007. *Unleashing India's Innovation – Toward Sustainable and Inclusive Growth*. Washington, DC: World Bank.

EARTO (European Association of Research and Technology Organizations). 2005. *Research and Technology Organizations in the Evolving European Research Area*. Brussels: EARTO.

Economy and Values Research Center. 2008. "National Competitiveness Report Armenia." Economy and Values Research Center, Yerevan.

El-Erian, Mohamed A., and A. Michael Spence. 2008. "Growth Strategies and Dynamics." *World Economics*, 9 (1): 57–96.

Engman, Michael. 2005. "The Economic Impact of Trade Facilitation." OECD Trade Policy Working Papers 21. Organization for Economic Co-operation and Development, Paris.

Eschenbach, Felix, and Bernard Hoekman. 2006. "Services Policy Reform and Economic Growth in Transition Economies. *Review of World Economics (Weltwirtschaftliches Archiv)*, 142 (4): 746-764.

European Commission. 2007. *Key Figures 2007: On Science, Technology and Innovation: towards a European Knowledge Area*. Brussels: European Commission.

European Commission. 2011. Innovation Union Competitiveness Report 2011. Luxembourg: Publications Office of the European Union.

Evenson, Robert E. 1984. "International Intention: Implications for Technology Market Analysis." In Zvi Griliches (ed.) *R&D, Patents, and Productivity*. Chicago: Chicago University Press.

Freeman, Christopher. 2006. "Catching Up and Innovation Systems: Implications for Eastern Europe." In Krzysztof Piech and Slavo Radosevic (eds.) *The Knowledge-Based Economy in Central and East European Countries: Countries and Industries in a Process of Change*. London: Palgrave Macmillan.

Gaidar, Yegor, Sergei Sinelnikov-Mourylev, and Nina Glavatskaya, eds. 2005. "Russian Economy in 2004. Trends and Perspectives." Issue 26. Moscow: Institute for the Economy in Transition.

Georghiou, Luke, Keith Smith, Otto Toivanen, and Pekka Yla-Anttila. 2003. *Evaluation of the Finnish Innovation Support System*. Helsinki: Ministry of Trade and Industry.

Gill, Indermit S., Fred Fluitman, and Amit Dar. 2000. *Vocational Education and Training Reform: Matching Skills to Markets and Budgets*. Cary, NC: Oxford University Press.

Goldberg, Itzhak, and Branko Radulovic. 2005. Hard Budget Constraints, Restructuring and Privatization in Serbia: A Strategy for Growth of the Enterprise Sector. Private Sector Note. World Bank, Washington, DC.

Goldberg, Itzhak, Lee Branstetter, John Gabriel Goddard and Smita Kuriakose. 2008. "Globalization and Technology Absorption in Europe and Central Asia: The Role of Trade, FDI and Cross-Border Knowledge Flows." World Bank Working Paper 150. World Bank, Washington, DC.

Goldberg, Itzhak, Manuel Trajtenberg, Adam Jaffee, Julie Sunderland, Thomas Muller and Enrique Blanco Armas. 2006. "Public Financial Support for Commercial Innovation: Europe and Central Asia Knowledge Economy Study Part I." *Europe and Central Asia Chief Economist's Regional Working Paper Series,* 1 (1), World Bank, Washington DC.

Gorodnichenko, Yuriy, Jan Svejnar, Katherine Terrell. 2010. "Globalization and Innovation in Emerging Markets." *American Economic Journal,* 2(2): 194–226.

Grossman, Gene M., and Elhanan Helpman. 1991. *Innovation and Growth in the Global Economy.* Cambridge, MA: MIT Press.

Hall, Bronwyn H., and John van Reenen. 2000. "How Effective are Fiscal Incentives for R&D? A Review of the Evidence." *Research Policy,* 29 (4-5): 449-470.

Helpman, Elhanan. 2004. "The Mystery of Economic Growth." *Journal of International Economics,* 68 (2): 518–27.

Helpman, Elhanan, and Manuel Trajtenberg. 1998. "Diffusion of General Purpose Technologies." In Elhanan Helpman (ed.), *General Purpose Technologies and Economic Growth.* Cambridge, MA: MIT Press.

Jaffe, Adam B. 1998. "The Importance of 'Spillovers' in the Policy Mission of the Advanced Technology Program." *Journal of Technology Transfer,* 23 (2): 11–19.

———. 2002. "Building Programme Evaluation into Design of Public Research-Support Programmes." *Oxford Review of Economic Policy,* 18 (1): 22–34.

Jaffe, Adam B., and Josh Lerner. 2001. "Reinventing Public R&D: Patent Law and Technology Transfer from Federal Laboratories." *Rand Journal of Economics,* 32: 167-198.

Jensen, Jesper, Thomas Rutherford, and David Tarr. 2007. "The Impact of Liberalizing Barriers to Foreign Direct Investment in Services: The Case of Russian Accession to the World Trade Organization." *Review of Development Economics,* 11 (3): 482–506.

Jones, Charles I., and John C. Williams. 1998. "Measuring the Social Return to R&D." *Quarterly Journal of Economics,* 113 (4): 1119–35.

Jones, Ronald, Henryk Kierzkowski, and Chen Lurong. 2005. "What Does Evidence Tell Us about Fragmentation and Outsourcing?" *International Review of Economics and Finance,* 14 (3): 305-16.

Keller, Wolfgang. 2002. "Trade and the Transmission of Technology." *Journal of Economic Growth,* 7 (1): 5–24.

Kinoshita, Yuko. 2000. "R&D and Technology Spillovers via FDI: Innovation and Absorptive Capacity." William Davidson Institute Working Paper 349. William Davidson Institute, University of Michigan Business School, Ann Arbor, MI.

Klette, Tor Jakob, Jarle Møen, and Zvi Griliches. 2000. "Do Subsidies to Commercial R&D Reduce Market Failures? Microeconomic Evaluation Studies." *Research Policy,* 29 (4-5): 471-495.

Krugman, Paul. 2000. "Technology, Trade, and Factor Prices." *Journal of International Economics,* 50 (1): 51-71.

Lach, Saul. 2002. "Do R&D Subsidies Stimulate or Displace Private R&D? Evidence from Israel." *Journal of Industrial Economics,* 50 (4): 369–90.

Lerner, Josh. 1999. "Small Businesses, Innovation, and Public Policy." In Zoltan J. Acs (ed.) *Are Small Firms Important? Their Role and Impact.* Boston: Kluwer Academic.

———. 2009. *Boulevard of Broken Dreams: Why Public Efforts to Boost Entrepreneurship and Venture Capital Have Failed—and What to Do About It.* Princeton, NJ: Princeton Press.

Lin, Justin Yifu, and Celestin Monga. 2010. "Growth Identification and Facilitation: The Role of the State in the Dynamics of Structural Change." Policy Research Working Paper 5313. World Bank, Washington, DC.

Lucas, Robert E., Jr. 1988. "On the Mechanics of Economic Development." *Journal of Monetary Economics,* 22 (1): 3–42.

Mattoo, Aaditya. 2005. *Economics and Law of Trade in Services.* Policy and Research Working Report. World Bank, Washington, DC.

Meske, Werner, ed. 2004. *From System Transformation to European Integration: Science and Technology in Former Socialist Central and East Europe at the Beginning of the 21st Century.* Piscataway, NJ: Transaction Publishers.

Middleton, John, Adrian Ziderman, and Arvil V. Adams. 1993. Skills for Productivity: Vocational Education and Training in Developing Countries. New York: Oxford University Press.

Modena, Vittorio, ed. 2002. *Israeli Financing Innovation Schemes for Europe: Final Report.* Pavia, Italy: University of Pavia.

Nastas, Thomas. 2005. *The Role of Venture Capital in European and Central Asian Countries.* Presented at the World Bank Knowledge Economy Forum IV, Istanbul, March 22–24.

Nelson, Richard, and Edmund Phelps. 1966. "Investment in Humans, Technological Diffusion, and Economic Growth." *American Economic Review,* 56 (2): 65–75.

Nicolaon, Gilbert. 2008. *The 'Restructuring' of Research and Development Institutes.* Presented at the 2008 Ukraine Knowledge Economy Training Initiative, Kiev, June 10.

OECD (Organisation for Economic Co-operation and Development). 2002. *Science, Technology and Industry Outlook 2002*. Paris: OECD.

————. 2003. *Governance of Public Research – Toward Better Practices*. Paris: OECD.

Pack, Howard, and Kamal Saggi. 2006. "The Case for Industrial Policy: A Critical Survey." *World Bank Research Observer,* 21 (2): 267–97.

Pack, Howard, and Larry E. Westphal. 1986. "Industrial Strategy and Technological Change." *Journal of Development Economics,* 22: 87–128.

Persson, Maria. 2008. "Trade Facilitation and the EU-ACP Economic Partnership Agreements." *Journal of Economic Integration,* 23 (3): 518-546.

Racine, Jean-Louis, Itzhak Goldberg, John Gabriel Goddard, Smita Kuriakose and Natasha Kapil. 2009. "Restructuring of Research and Development Institutes in Europe and Central Asia." World Bank, Washington, DC.

Radosevic, Slavo. 1998. "The Transformation of National Systems of Innovation in Eastern Europe: Between Restructuring and Erosion." *Industrial and Corporate Change,* 7 (1): 77–108.

————. 2005. *Are Systems of Innovation in Central and Eastern Europe Inefficient?* Presented at the DRUID Tenth Anniversary Summer Conference 2005 on Dynamics of Industry and Innovation: Organizations, Networks and Systems, Copenhagen, Denmark, June 27–29.

Radosevic, Slavo, Maja Savic, and Richard Woodward. 2010. "Knowledge based Entrepreneurship in CEE: Results of a Firm Level Survey." In Franco Malerba (ed.) *Knowledge Intensive Entrepreneurship and Innovation Systems Evidence from Europe*. London, UK: Routledge.

Republic of Serbia Privatization Agency. 2005. *Impact Assessment of Privatisation in Serbia*. Belgrade: Republic of Serbia Privatization Agency.

Rodrik, Dani. 2004. "Industrial Policy for the Twenty-First Century." CEPR Discussion Paper 4767, Centre for Economic Policy Research, London.

Romer, Paul M. 1986. "Increasing Returns and Long-Run Growth." *Journal of Political Economy,* 94 (5): 1002–37.

————. 1990. "Human Capital and Growth: Theory and Evidence." *Carnegie-Rochester Conference Series on Public Policy,* 32 (1): 251–86.

Rothschild, Lord. 1971. "Organization and Management of Government Research and Development." In *A Framework for Government Research and Development*. London: Her Majesty's Stationery Office.

Rutherford, Thomas, David Tarr and Oleksandr Shepotylo. 2007. "The Impact on Russia of WTO Accession and The Doha Agenda: the Importance of Liberalization of Barriers against Foreign Direct Investment in Services for Growth and Poverty Reduction," In L. Alan Winters (ed.) *The WTO and Poverty and Inequality*. Cheltenham, UK: Edgar Elgar Publishing.

Schaffer, Mark E., and Boris Kuznetsov. 2008. "Productivity." In Raj M. Desai and Itzhak Goldberg (eds.) *Can Russia Compete? Enhancing Productivity and Innovation in a Globalizing World.* Washington, DC: Brookings Press.

Sedaitis, Judith. 2000. "Technology Transfer in Transitional Economies: A Test of Market State and Organizational Models." *Research Policy,* 29 (2): 135–47.

Shefer, Daniel, and Amnon Frenkel. 2002. "An Evaluation of the Israeli Technological Incubators Program and Its Projects, Final Report." The S. Neaman Institute for Advanced Studied in Science and Technology, Technion, Haifa, Israel.

Tan, Hong, Vladimir Gimpelson, and Yevgeniya Savchenko. 2008. "Upgrading Skills." In Raj M. Desai and Itzhak Goldberg (eds.) *Can Russia Compete? Enhancing Productivity and Innovation in a Globalizing World.* Washington, DC: Brookings Press.

Teubal, Morris, and Gil Avnimelech. 2002. "Israel's Venture Capital (VC) Industry: Emergence, Operation and Impact." Jerusalem Institute for Israel Studies, Jerusalem.

Toren, Beny.1900. "R&D in Industry." In David Brodet, Moshe Justman, and Morris Teubal (eds.) *Industrial Technological Policy for Israel.* Jerusalem: The Jerusalem Institute for Israeli Studies.

Trajtenberg, Manuel. 2001. "R&D Policy in Israel: An Overview and Reassessment." In Maryann P. Feldman and Albert N. Link (eds.) *Innovation Policy in the Knowledge-Based Economy.* Boston: Kluwer Academic Publishers.

Trefler, Daniel, and Diego Puga. 2010. "Wake Up and Smell the Ginseng: International Trade and the Rise of Incremental Innovation in Low-Wage Countries." *Journal of Development Economics,* 91(1): 64-76.

Veugelers, Reinhilde, Karl Aiginger, Dan Breznitz, Charles Edquist, Gordon Murray, Gianmarco Ottaviano, Ari Hyytinen, Aki Kangasharju, Mikko Ketokivi, Terttu Luukkonen, Mika Maliranta, Markku Maula, Paavo Okko, Petri Rouvinen, Markku Sotarauta, Tanja Tanayama, Otto Toivanen, and Pekka Ylä-Anttila. 2009. *Evaluation of the Finnish National Innovation System – Policy Report.* Helsinki: Taloustieto Oy.

Wakasugi, Ryuhei. 1986. "Gijutsu kakushin to kenkyu kaihatsu no keizai bunseki: Nihon no kigyo kodo to sangyo seisaku" (The economic analysis of research and development and technological progress: Japanese firm activity and industrial policy). Tokyo: Toyo Keizai Shimposha.

———. 1990. "A Consideration of Innovative Organization: Joint R&D of Japanese Firms." In Arnold Heertje and Mark Perlman (eds.) *Evolving Technology and Market Structure: Studies in Schumpeterian Economics.* Ann Arbor, MI: The University of Michigan Press.

Wallsten, Scott J. 2000. "The Effects of Government-Industry R&D Programs on Private R&D: The Case of the Small Business Innovation Research Program." *RAND Journal of Economics,* 31 (1): 82–100.

Wilson, John S., Catherine L. Mann, and Tsunehiro Otsuki. 2003. "Trade Facilitation and Economic Development: A New Approach to Quantifying the Impact." *World Bank Economic Review,* 17 (3): 367–89.

World Bank. 2005a. *From Disintegration to Reintegration: Eastern Europe and Former Soviet Union in International Trade.* Washington, DC: World Bank.

World Bank. 2005b. *From Transition To Development: A Country Economic Memorandum for the Russian Federation.* Washington, DC: World Bank.

World Bank. 2007. "Implementation Completion and Results Report (IDA-35800) on a Learning and Innovation Loan (LIL)." World Bank, Washington, DC.

World Bank. 2010a. "E-Society and Innovation for Competitiveness (EIC) Project." Project Appraisal Document. World Bank, Washington, DC.

World Bank. 2010b. "Implementation Completion and Results Report: Renewable Energy Project to the Republic of Turkey." World Bank, Washington, DC.

World Bank. 2011a. Europe 2020 Poland: Fueling Growth and Competitiveness in Poland through Employment, Skills and Innovation. Warsaw: Protea-Taff Studio DTP.

World Bank. 2011b. "Poland Enterprise Innovation Review." World Bank, Washington, DC.

World Bank. 2011c. "Sustaining Reforms under the Oil Windfall." Russian Economic Report 25. World Bank, Washington D.C. and Moscow. All reports available at: http://www.worldbank.org/eca/rer.

World Economic Forum. 2010. Global Competitiveness Report 2009–2010. Geneva: World Economic Forum.